D0446440
THE
RESTLESS
NORTHWEST
A GEOLOGICAL STORY

THE RESTLESS NORTHWEST

A GEOLOGICAL STORY

Hill Williams

Washington State University Press
Pullman, Washington

Washington State University Press
PO Box 645910
Pullman, Washington 99164-5910
Phone: 800-354-7360 Fax: 509-335-8568
E-mail: wsupress@wsu.edu
Web site: www.wsu.edu/wsupress

First printing 2002

Printed and bound in the United States of America on pH neutral, acid-free paper.

ISBN 0-87422-250-8 (paperback)

Table of Contents

❖❖❖

To the memory of my parents,
Hill and Ursula Williams,
who would have been fascinated by
this story of the country they loved.

❖❖❖

Acknowledgments

Putting this book together required a lot of patience, but not necessarily on my part. Over the years, as I wrote science news for the *Seattle Times* and later worked on this book, I was fortunate to meet scientists who were willing to explain what they were doing and why it mattered. As I listened, I was usually trying to keep up in the mountains, scrambling on steep coulee walls, accompanying them on research ships, or trekking through a tunnel taking us millions of years back in time through an ancient lava flow.

Almost without exception they took time from their busy schedules to help me understand the geology of the Pacific Northwest. They had to be patient because I lack a formal scientific background. But in a way that's good. I always figured if I could get something through my head, I could do a better job of telling it to other nonscientists. That's what I've tried to do here. If you enjoy this book, I hope you'll give a silent thank you to the scientists who did the work—and took the time to explain it to me.

After all the years and all the miles, I've probably forgotten some names, and I apologize in advance. But here are at least most of those who helped me:

Darrell S. Cowan, Eric Cheney, Joseph A. Vance, Stanley E. Chernicoff, Stephen C. Porter, Estella B. Leopold, Andy Moore, Stephen D. Malone, H. Paul Johnson, John Delaney, Gary D. Webster, Peter Hooper, Eric V. McDonald, Alan J. Busacca, Myrl E. Beck, Jr., David C. Engebretsen, Edwin H. Brown, Dale Stradling, Marvin Beeson, John Eliot Allen, Paul Hammond, Laverne Kulm, Robert A. Duncan, Paul Komar, Steve Reidel, Robert J. Carson, Pete Sak, Donald A. Swanson, Ralph

Haugerud, Richard B. Waitt, Jr., Craig S. Weaver, Robert Powell, James W. Whipple, Brian Atwater, Rowland Tabor, Robert D. Bentley, Robin Riddihough, Don Tubbs, Keith Stoffel, Weldon Rau, A. Lincoln Washburn.

Glen Lindeman at Washington State University Press lent early encouragement. Louise Freeman-Toole, my editor, made many good suggestions. Floyd Bardsley used his computer magic to make my sketches look professional.

And, most importantly, I thank my wife Mary Lou, who helped make the whole project fun.

Pacific Northwest Geological Timeline

Our 24-hour clock starts ticking. It is 12:01 A.M.

175 million years ago:	A supercontinent breaks apart, pushing North America west and Europe east, triggering assembly of the Pacific Northwest, half a planet away.
	EUROPE AND NORTH AMERICA SEPARATE
100 million years ago:	10:17 A.M. Building blocks collide with North America's advancing western edge.
	DINOSAURS GO EXTINCT
50 million years ago:	5:08 P.M. Parts of the Pacific Northwest are in place, but most are still under water.
	MAMMALS DOMINATE
40 million years ago:	6:30 P.M. Undersea volcanoes deposit foundations for the Cascade and Coast ranges.
	ROCKY MOUNTAINS RISE
17 million years ago:	9:40 P.M. Massive outpourings of lava flood a swampy basin east of the growing Cascades.
	PRE-HUMANS APPEAR
9 million years ago:	10:46 P.M. Volcanoes roar back to life, resume building the Cascades.
	STONE TOOLS INVENTED
500,000 years ago:	11:55 P.M. Mount Rainier is born.
	FIRE TAMED
40,000 years ago:	11:59 P.M. Mount St. Helens is born.
	HUMANS ENTER NORTH AMERICA
14,000 years ago:	11:59:36 P.M. Great floods rip across the Columbia Basin as the ice age dies.
10,000 years ago:	The Pacific Northwest looks much as we know it.

Part I

The Ancient Northwest

A Puzzle Assembled from Around the World

A dramatic undersea terrain still being created off the Pacific Northwest coast illustrates how our region was formed millions of years ago.

1

A Mirror Image Offshore

We of the Pacific Northwest live in a patchwork land pieced together of odd fragments that drifted here from somewhere else. It's a crazy quilt made of distinct groups of rocks, each with a different history, separated by ancient faults. We drive our cars across cracks in the earth's surface that once wrenched the ground in powerful earthquakes. We cross mammoth flows of lava that cooled thousands of years ago, and picnic on ocean beaches that have been squeezed so severely they turned on edge or even overturned long ago. We hike mountains that have been uplifted a mile over millions of years and are still being squeezed higher. Our city foundations rest on volcanic rock that erupted beneath an ancient sea. All of the Pacific Northwest underwent this piecemeal construction except for Idaho and the very eastern edge of Washington, which are part of the original continent.

A Gargantuan Assembly Job

Our corner of the country is the end result of a gargantuan assembly job involving forces of unimaginable proportions squeezing, cracking, pulling, ripping. Did we say end result? Not on your life!

The landscape we know—deserts, mountains, volcanoes, rocky headlands jutting into the ocean—is still being pushed and twisted and compressed.

Oh, sure, we know we have a live volcano in the neighborhood, and we remember the earthquakes that scare us every few years. But the soothing calm we sometimes feel gazing at beloved ocean beaches, desert sunsets, lakes and rivers, farms, or mountains reaching toward the sky could easily be shattered.

In fact, the comfort zone of geologists, building engineers, and some of the public took a serious hit in the 1990s. There was surprising new evidence of great earthquakes in which an island beach jumped 20 feet, coastal marshes dropped below sea level, and a great wave raced in Puget Sound. Even as you read this, relentless pressure driving onshore is tilting up the outer coasts of Oregon and Washington, pushing down areas behind the Coast Range—a little like a huge teeter-totter. Coastal towns are inching toward the northeast. Mountains are being squeezed closer together, imperceptibly but measurably. The push is gradual, relentless, irresistible. A sudden break, a sudden relaxing of pressure, has caused earthquakes in the past greater than any recorded in modern decades. And anything that has happened before is almost certain to happen again. Sometime.

What's putting the squeeze on the Northwest?

To find the source of the pressure exerted upon this region, we must look offshore beyond the spectacular coastlines of Oregon, Washington, and British Columbia. This patch of ocean floor, called the Cascadia basin, is little known to most of us who live right next door. But it played an important role in assembling the Pacific Northwest. The forces generated as the Cascadia basin and North America creep toward and collide with each other will continue to shape our corner of the country for millennia to come.

The undersea terrain is as dramatic as anything on dry land: broad plains traversed by rugged canyons and bounded by mountain ranges with active volcanoes. But the canyons and volcanoes are in total darkness, covered by water more than a mile deep, far below where sunlight penetrates. Only a few exploring oceanographers have caught a glimpse of these features; even these scientists haven't seen beyond the range of spotlights on their research submarines.

The Cascadia Basin

Roughly the size and shape of Oregon and Washington, the basin and its bordering undersea mountain range are like a mirror image of the Pacific Northwest we know on land. So the Cascadia basin, together with the nearby Juan de Fuca ridge, make an appropriate place to begin the story of how the Pacific Northwest was put together, why it looks as it does, and why scientists come from all over the world to study it.

The basin, mostly covered with a thick layer of sediment, is separated from the much deeper Pacific Ocean bottom by the Juan de Fuca ridge, an underwater range of volcanic mountains. The ridge is about as bulky as the Cascade Range and is generally parallel to the coastline, although with its zigs and zags it varies from about 150 to 300 miles offshore. (For a rough idea of distances, the ridge is about as far west of the mouth of the Columbia River as the Idaho border is to the east.)

Although oceanographers had known from depth soundings that the ridge was there, the mountains had never been seen by humans until the 1980s when deep-diving research submarines first probed the inky black depths. Crowded into the little submarines, fascinated scientists peered through thick, plexiglass windows as the sub's searchlight probed the strange world.

Hot-water geysers gushed from the ocean bottom. There were fanciful structures—the explorers called them "castles"—built by minerals precipitated from the superheated water of the geysers where they encountered the near-freezing ocean-bottom water. One structure grew 2 inches in one day; another monitored by a video camera grew 7 feet in 6 weeks.

There was even a "waterfall" that fell up, caused by extremely hot water trying to rise through cold water. Because the hot water had different reflective properties than cold water, it looked like a mirror. Bemused scientists watched as drops of reflecting hot water formed a puddle beneath a horizontal ledge of a "castle," filled until it overflowed and then "fell" up to the next level.

The geysers are dangerously hot: the tough fiberglass skin of a submarine blistered when the pilot inadvertently backed into a geyser; and a piece of thick plastic rope melted when it was pulled through a hot-water jet. The water, at 400 degrees Fahrenheit, would flash into steam if not for the pressure of a mile of water above it.

The Juan de Fuca Ridge

The Juan de Fuca ridge is part of a system of interconnected undersea mountain ranges that occur over rifts or weak places in the ocean bottom, allowing molten rock to rise from the earth's interior. As the molten rock is squeezed from the ocean floor and encounters cold seawater, it cools suddenly and solidifies into a rock known as basalt. Over millions of years, the accumulating basalt in the Cascadia basin built a small ridge that grew into a bigger one and finally into a full-size mountain range—the Juan de Fuca ridge. As the mountains got higher and their slopes steeper, a giant slab of rock began to slip, just as happens on dry land when slopes become unstable. (Although other factors may be at work, some geologists believe gravity is the primary moving force.) The immensity is difficult to imagine: it's as though the entire eastern and western slopes of the Cascades, fed by oozing lava along the crest, were sliding downhill.

The Juan de Fuca Plate

The basalt slab sliding eastward from the ridge becomes new ocean bottom as it reaches the foot of the ridge and continues its slow movement toward the coastline. It is now the Juan de Fuca plate, which plays an important role in our story of how the Pacific Northwest was assembled. Its twin, the gigantic slab sliding west

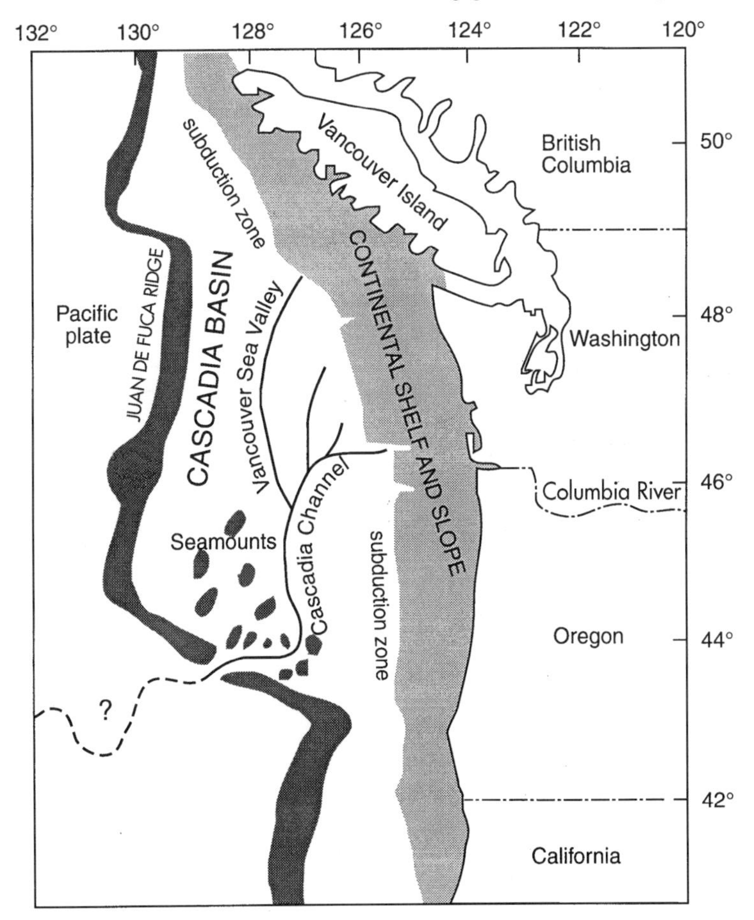

The Juan de Fuca ridge encloses Cascadia basin and its deep-sea channels.

toward the deep ocean, becomes part of the immense northwest-creeping Pacific plate that forms much of the Pacific Ocean floor. As these sections of once-molten rock move away from the ridge, they continue to cool, thicken, and sink farther beneath the sea surface.

The Juan de Fuca plate creeps eastward toward North America about an inch a year, or about as fast as fingernails grow. In a human lifespan, this ponderous, miles-thick chunk of rock would move about 6 feet. The pace is so slow that the "new" ocean bottom is about 10 million years old when it reaches the foot of the continental slope, the true edge of North America. At that point, the heavier ocean plate slides beneath the lighter continental plate; over millions of years the crumpling and twisting caused by this collision helped create the spectacular landscape of the Pacific Northwest.

A View of Life's Origins

The exploration of deep-sea volcanic vents discovered in the eastern Pacific Ocean in the late 1970s and early 1980s yielded many surprises. Previously, biologists had believed all life was based on the photosynthetic process, in which life forms are energized by sunlight. But there—in water more than a mile deep, far deeper than sunlight could reach—were clams, mussels, barnacles, bacteria, and other life forms that apparently derive nutrients from chemicals in the heated water, chiefly hydrogen sulfide. Among them were tube worms (the name pretty well describes their structure) up to nine feet long. Instead of burrowing into soft sediment, these creatures were clinging to hard rock, either basalt or the mineral structures.

Some of the life forms clustered around active volcanic vents in the deep sea were later found to have a primitive genetic makeup, much different from their modern counterparts on land. It has prompted some to wonder if life on earth originated in the hot, sulfurous cracks in the ocean floor.

Incredible Cities and Hanging Gardens

If plates are spreading in opposite directions from the ridge, there should be a gap between them as they separate along the crest of the ridge. The gap should be the hottest spot because it's right above the rising magma. That's exactly what instruments found: a narrow, steep-sided valley along the ridge crest with superheated geysers spewing into the floor of the valley. In some places, instruments even recorded a narrow cleft in the floor of the valley, marking the exact boundary between the Pacific and Juan de Fuca plates. It was just the way theory said it should be, which may have been the biggest thrill for exploring scientists—even more, perhaps, than seeing the ornate "castles" and superhot geysers.

Scientists believe deep-sea geysers are created as magma forces its way upward and fractures the old ocean floor, allowing seawater to seep into deep cracks. Water approaching the very hot rocks near the magma heats up and begins dissolving minerals from the rock. The hot water, now buoyant, rises through the fractured rock of the ocean bottom and, under great pressure, gushes upward as geysers. Seafloor water is only a few degrees above freezing and it quickly chills the geyser water. As the water temperature drops hundreds of degrees in a few seconds, the water loses its ability to hold minerals in solution; the minerals are deposited on the ocean floor, eventually building a structure around the geyser.

The ocean-bottom rift beneath the Juan de Fuca ridge continues thousands of miles south. The first geysers spouting from mineral structures were discovered off South America in the 1970s. Oceanographers called them "chimneys"; if black, mineral-laden hot water was gushing from the top, they were called "smokers."

As the research submarines probed the Juan de Fuca ridge in the 1980s, scientists discovered that its mineral structures are even bigger and more ornate than those off South America. The sight prompted usually conservative scientists to refer to castles and clusters of structures as "incredible cities." One neighborhood of castles off Vancouver Island, dubbed "Magic Mountain" by the explorers, extends for more than 150 yards. Research submarines are dwarfed by the structures, some of which are 30 to 60 feet high. The submarines' lights revealed colonies of huge tube worms hanging from projecting shelves on the castles; one exploring scientist said the scene looked like hanging gardens.

Direct human observation of the undersea world is still limited to what little a submarine's light can illuminate. In fact, one of the exploring scientists, John R. Delaney of the University of Washington, said: "It's as hard to describe the environment down there as it would be for you to describe the Cascade Mountains to a person who'd never been out of Kansas, and you were basing your description on a trip to the mountains at night with a flashlight."

A Volcano Being Pulled Apart

The Juan de Fuca ridge includes a volcano unfortunate enough to straddle the boundary between the Pacific and Juan de Fuca plates; as a result, it is being pulled apart. Axial volcano, almost due west of the mouth of the Columbia River, is so named because it sits astride the axis of the seafloor ridge as the two plates pull apart.

Axial volcano is about 12 miles in diameter and rises almost a mile from the ocean floor. Its crater is a huge collapsed area, properly called a caldera, about 6 miles long and 2 miles wide. Geologists assume that the floor of Axial's caldera, like those of similar volcanoes on the island of Hawaii, rises as magma is pumped into the mountain and then falls as the eruption begins. The floor may rise and fall as much as several hundred feet. Pressure from beneath may force open cracks in the caldera floor, allowing magma to seep out. In the 1980s scientists found basalt in Axial's caldera estimated to be less than 10 years old.

Megaplumes

Volcanic eruptions along the Juan de Fuca ridge off the Oregon coast (detected by land-based instruments) in the mid-1990s served as a reminder that the area is still active. And in the mid-1980s something even bigger must have happened along the undersea mountain range—an event discovered more or less by accident by oceanographers studying how ocean-bottom geysers affect the composition of the surrounding seawater.

Oceanographers aboard deep-diving submarines had previously studied the plumes of hot water emitted by geysers on Juan de Fuca ridge. The plumes rise and then drift away on currents, much like smoke blowing away from smokestacks on land; they typically reach a height of several hundred feet. This time, ship-towed instruments detected a megaplume drifting 3,000 feet above the Juan de Fuca ridge but still half a mile below the ocean surface. Not only was it huge, but it retained its shape for days as the ship crisscrossed it time and again to map its size, take its temperature, and analyze its chemical makeup.

The disc-shaped megaplume was about 16 miles in diameter and up to 2,300 feet thick. Its temperature and load of chemicals were 5 to 10 times greater than those found in the smaller plumes just above the ridge. But, significantly, the minerals, although more concentrated, were the same as those found in ocean-bottom geysers and in smaller plumes.

Although they couldn't prove how or where the megaplume formed, the oceanographers were almost certain it represented an enormous amount of fluids released over a very short time—a matter of days—from the ocean-bottom mountain range. It could have been caused by an earthquake that cracked open a reservoir of hot water that normally leaked out slowly and steadily from the rock. Or it could have been a sudden change in pressure in the reservoir that blew vast quantities of the mineralized fluid into the ocean water. Axial was considered a possible source of the megaplume, although the volcano was 60 miles away.

The group of scientists went back the next summer to see if the megaplume was still there. Instead, they found another plume, somewhat smaller than the first but still surprisingly large. It was about 25 miles north of their earlier discovery, and like the first, drifted over the Juan de Fuca ridge. It was an important discovery because it strengthened evidence that seafloor spreading centers, in addition to creating new earth crust, play an important role in the chemical composition of the oceans, both through slow leaks of fluids and perhaps by occasional enormous burps.

A Shower of Sediment

Newly formed basalt on the upper flanks of the Juan de Fuca ridge is almost bare rock. But during the millions of years the plate creeps toward the continent, it is showered with sediment. Millions of years later when rock reaches the foot of North America's continental slope, the blanket of sediment ranges from 2 to 2½ miles thick.

There are two sources of this sediment: skeletons of tiny marine animals that fall slowly from the ocean's life zones far above; and, more importantly, sand and clay eroded off the continent and carried out to sea by rivers and streams. The Cascadia basin traps

Icebergs Off the Northwest Coast

There are clues in Cascadia basin that icebergs once floated off Washington and Oregon. Oregon State University oceanographers found pebbles embedded in clay in deep water far out in the basin. The mixture of pebbles in clay was puzzling because when flowing water carries sediment, the heavier material drops out first as the current lessens: rocks are deposited at the bottom, pebbles in the middle, and small particles of clay on top.

The mineral signature of the pebbles indicated they were eroded from rocks in the Puget Sound basin and adjacent mountains in Washington and British Columbia. The oceanographers concluded that the pebbles must have rafted south on icebergs that broke away from glaciers near the mouth of the Strait of Juan de Fuca. Ocean currents carried the glacial fragments south along the coast. As the icebergs melted, the jumble of silt, pebbles, and boulders caught in the ice dropped all in a heap—not in neat layers like water-carried debris. Ice-rafted pebbles were found as far south as 44 degrees north latitude—about 30 miles south of modern-day Newport, Oregon—apparently as far as icebergs traveled before they melted.

the vast amount of sediment pouring off the continent to form one of the thickest sediment layers in the entire Pacific Ocean.

Judging from depth measurements, much of the Cascadia basin must resemble a plain of gently rolling hills where the slightest disturbance would raise blinding clouds of sediment that take hours to clear. The plain is cut by large sea valleys that rival those on land in both size and length. The longest is the Cascadia channel, which is longer than the Columbia River.

The Cascadia Channel

Stretching for 1,300 miles, the Cascadia channel is the longest, deep-sea channel known in the Pacific Ocean. This great underwater canyon and others like it in the Cascadia basin are a product of the same processes that dug great canyons on land such as Grand Coulee and Moses Coulee in central Washington. These now-dry coulees were carved by rivers, and, in the late ice age, by flood waters loaded with rocks, chunks of ice, gravel, sand, clay—all of which increased the water's cutting ability.

Much the same process carved the Cascadia channel, with one difference: the coulees were dug by roaring, tumbling, fresh water, while the seafloor channels were carved by silent, bottom-hugging torrents of seawater heavy with particles of sand and clay. These currents, called turbidity currents, originate on the continental slope where rivers have delivered huge loads of sediment. When the slope becomes too steep, the whole pile begins to slide, triggering a turbidity current that races and carves its way down the slope and across the basin.

Turbidity currents gouged out the Cascadia channel about 3 million years ago. The channel was cut deeper into the seafloor during the ice age as fast-flowing streams carried heavy loads of sand and gravel toward the ocean. Sea level was more than 300 feet lower during the ice age and rivers dug canyons across the exposed continental shelf as they headed for the ocean. Those canyons are now deep under water and it is at the foot of one, about 25 miles west of Willapa Bay on the Washington coast, that the Cascadia channel begins its long run.

Where the channel heads west, it has built levees as high as 90 feet above the surrounding plain. About 100 miles offshore, the channel turns south to run between chains of seamounts—extinct volcanoes, which stand like ghostly sentries on the ocean bottom. The Cascadia channel finally leaves the basin through a ridge of fractured rock, the Blanco fracture zone, off the southern coast of Oregon. The zone accommodates differing plate movements on either side by sliding, rumpling, and fracturing. Here, as the Cascadia channel passes through the disrupted zone, it changes from U-shape to a V, with the channel floor narrowing to only 20 or 30 feet wide. Seamounts forming the canyon walls tower as much as 3,000 feet above the channel floor. As the channel emerges from the rumpled fracture zone, it falls hundreds of feet and spreads out on the deep floor of the ocean.

The Cascadia Subduction Zone

Just as the Cascadia basin is created by volcanoes along the Juan de Fuca ridge, it is swallowed up some 300 miles away at its eastern boundary, which helps explain why the earth is not expanding as new seafloor is created. As the Juan de Fuca plate approaches North America, it collides with the foot of the continental slope, miles offshore and deep underwater. At this point, the Cascadia subduction zone, the plate bends more sharply downward and dives, or subducts, beneath the edge of the continent.

The Cascadia subduction zone is unique among similar features around the world. Unlike other subduction zones, there is no ocean-bottom trench formed by the subducting plate bending down to begin its dive. The trench may have been filled with the enormous amount of sediment produced by ice-age glaciers and delivered to the ocean by rivers and streams throughout the region, primarily the Columbia between Oregon and Washington and the Fraser River in British Columbia.

As the Juan de Fuca plate squeezes beneath the continent, much of its miles-thick layer of sediment gets scraped off and sticks to the continental slope. Visualize trying to slide a sheet of cardboard covered with peanut butter underneath another piece of cardboard. Much of the peanut butter would scrape off on the upper sheet. Something like that happens in the subduction zone. The ongoing collision has pushed sediment up onto the continental slope, forming a series of ridges parallel to the coastline.

Exploring scientists first saw the ridges in the mid-1980s, viewing them from research submarines. The bottom of the Cascadia basin (the sediment-covered Juan de Fuca plate) is relatively flat on the ocean side of the subduction zone. But in the zone of collision with the continent, the bottom of the basin becomes rumpled and deformed into ridges. One oceanographer said the effect is as though a wedge had been driven beneath another wedge and then another wedge driven under both of them and so on to produce a series of parallel ridges.

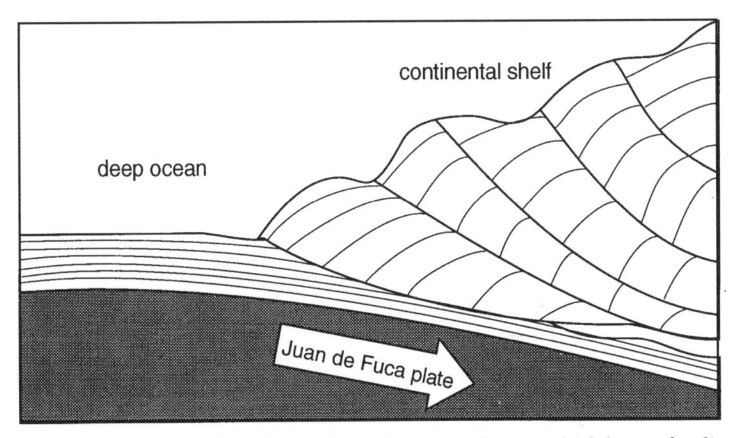

As the Juan de Fuca plate thrusts beneath the continent, a thick layer of sediment gets scraped off, wedging up ridges at the foot of the continental slope.

Oceanographers, studying the subduction zone from a submarine, observed a remarkable sight: water oozing from sediments under compression, like water being squeezed out of a sponge. Although the water was cold (not at all like the hot-water geysers on the volcanic Juan de Fuca ridge), it was still able to support big clams and tube worms. These colonies, 9,000 feet below the surface, apparently live on organic material produced by bacteria, which in turn utilize hydrogen sulfide squeezed from the compressed sediments.

The observation of water actually being squeezed from sediment was striking evidence of the tremendous energy produced by the colliding plates, energy that continues to shape the Pacific Northwest.

❖❖❖

Puzzling "stripes" on the ocean bottom lead to the discovery of how continents move.

2

Beneath a Mile of Water

Evidence that came together in the 1990s was a real wake-up call to scientists who study earthquakes in the Northwest, as well as to those of us who work and live in the region. A number of research projects pointed to an important conclusion: There have been great earthquakes in the Pacific Northwest's past, more powerful and destructive than any recorded in recent times. And there is a pretty good chance of a repeat, maybe tonight, maybe next year, maybe next century. But almost certainly sometime. Such an earthquake in today's Pacific Northwest would take a devastating toll in death and damage.

The bits of information that made possible the sobering research results were a long time coming. In fact, you might say they began in a mystery hidden beneath more than a mile of water. This is how it happened:

It was soon after World War II, heady years when scientists released from military service or wartime work had an assortment of new scientific toys to play with, instruments developed during the war and now being converted to research. Arthur D. Raff and Ronald G. Mason of the Scripps Institution of Oceanography, La Jolla, California, were among researchers who took some of the newly available instruments to sea in the early 1950s. Raff and

Mason were aboard a research ship towing a magnetometer off the Oregon-Washington-British Columbia coast, part of a study of the ocean bottom. As the magnetometer moved along near the sea floor, it detected faint magnetic forces in the rock beneath the sediment layer. The impulses were conveyed through more than a mile of wire to the mother ship, the *Pioneer,* where they were plotted on a chart.

Scientists had known for many years that volcanic rock carries a record of the earth's magnetic field at the time the rock crystallized as it cooled from the molten state. And they also knew that the magnetic field, for reasons that are not clear, has reversed itself repeatedly in the distant past. In other words, compass needles that now point to magnetic north at times would have pointed to magnetic south. We call the present magnetic field "normal" because it's been that way since the last reversal 700,000 years ago.

Raff and Mason were not prepared for the unexpected pattern that appeared on their chart. They found that "normally" magnetized rock in the ocean floor alternated with rock of reversed magnetism in long stripes. Furthermore, these alternating stripes ran north-south, roughly parallel to an undersea mountain range several hundred miles off the coast called the Juan de Fuca ridge.

Raff and Mason proceeded cautiously. They delayed reporting their findings until they had checked and rechecked the numbers and satisfied themselves that their instruments hadn't been playing tricks. Finally, in the 1961 journal article in which they reported their findings, they concluded: "There is as yet no satisfactory explanation." The discovery surprised and puzzled other researchers who also were at a loss to explain the strange stripes.

A Global Solution to a Geological Puzzle

Within a couple of years after the Raff-Mason article appeared, the mystery began to be unraveled. Oceanographers were discovering that the undersea mountains off the Pacific Northwest are

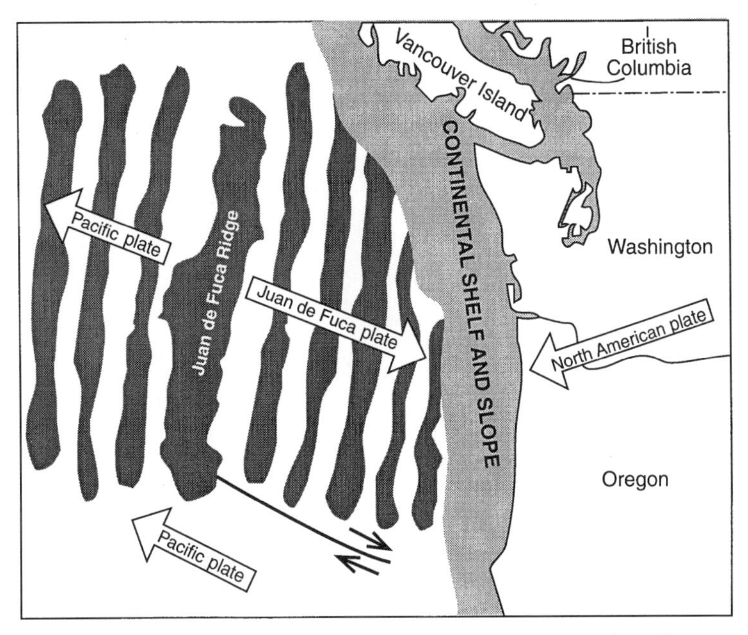

Differently magnetized "stripes" of ocean bottom moving away from the Juan de Fuca ridge puzzled oceanographers at the time of their discovery in the 1950s.

part of a chain of ocean-bottom mountain ranges girdling the globe like the seams on a baseball. Harry H. Hess of Princeton University suggested that if the undersea mountains were formed by magma oozing up from the earth's interior, the molten rock would create new areas of ocean floor, which would then move sideways away from the mountain ranges.

If that happened, rocks farthest from the ridge should be older than those nearest the ridge, thus suggesting a way to test the idea. Building on Hess's idea, British scientists Frederick J. Vine and Drummond H. Matthews proposed that "if spreading of the ocean floor occurs, blocks of alternately normal and reversely magnetized material would drift away from the centre of the ridge and parallel to the crest of it." This pattern of magnetism, they said,

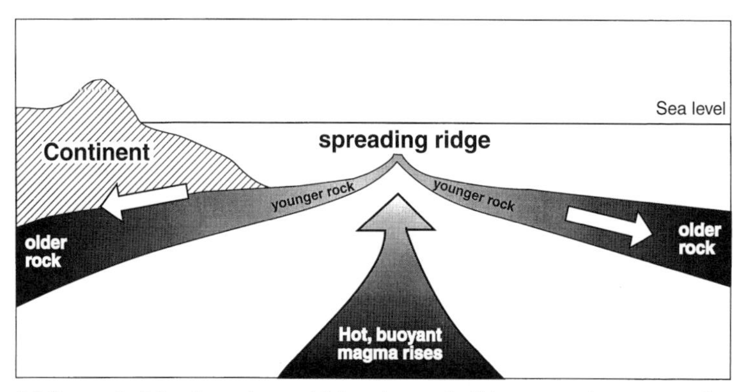

Molten rock rising from the earth's interior can build a mountain range on the ocean floor. As newly formed ocean bottom slides away from the ridge it may carry a continent with it.

could explain the mysterious stripes documented on the seafloor off the Pacific Northwest.

The next step was to see if the magnetic stripes on the ocean floor could be matched with the known ages of magnetic-field reversals in rocks on land. It turned out that Hess's theory was correct: Rock farthest from the undersea mountain ranges was oldest and that closest to the ridges the youngest.

Scientists began referring to the undersea ridges as spreading centers. By knowing the age of a section of ocean bottom and its distance from the spreading center, it was possible to calculate its average speed. Some sections of the seafloor moved at the surprising speed of four inches a year.

The realization that the ocean floor moves gave credence to suggestions dating from early in the twentieth century that modern continents formerly had been joined and then had drifted apart. Geologists had long puzzled over the apparent "fit" of South America and Africa and the belts of identical plant and animal fossils found on continents now separated by ocean. German scientist Alfred Wegener proposed in 1912 that the continents had once been joined; he even drew a map fitting today's continents into an ancient supercontinent. But for the first half of the century, most geologists dismissed the idea; they couldn't understand how continents could plow through the solid rock of the earth's crust. It wasn't until the 1960s that it became clear to geologists that continents don't need to plow through rock: they ride piggyback aboard huge plates of oceanic crust that move away from spreading centers.

A World Open to Reinterpretation

Those were exciting years for oceanographers and geologists, especially so for Tanya Atwater, a young graduate student at the Scripps Institution of Oceanography, home base for Raff and Mason who had first detected the magnetic stripes. On a big chart in his Scripps laboratory, H.W. "Bill" Menard had plotted all the magnetic measurements collected to that point by ships steaming across the Pacific.

"It seemed that the whole world lay waiting for reinterpretation," Atwater wrote years later. "Although I was not officially working with Bill, I was almost irresistibly drawn to his laboratory and to this map whenever I had a spare moment. The sea floor-spreading story of the region unfolded before our eyes. He was like a child in a candy store; I was in heaven." Atwater published a landmark scientific article in 1970 that for the first time correlated movements of the ocean plates with the geologic evolution of the West Coast over millions of years.

Even before Atwater's 1970 article, J. Tuzo Wilson, a Canadian geophysicist, had

How Plates Move

The earth's crust (the solid surface on which we live) averages about 30 miles thick beneath the continents, thinner beneath the oceans. The crust is the surface layer of the 60-mile-thick lithosphere which floats on the hot, ductile asthenosphere. Huge ocean plates, some of them carrying continents, can slide on the asthenosphere, something like sheets of cardboard moving on very thick syrup. The driving force is believed to be convection currents created by intense heat deep in the earth's interior.

become a convert. Wilson had opposed earlier theories of continental drift but the idea of seafloor spreading caused him to change his mind abruptly in the 1960s. Although he did little of the actual research, he is credited with being the great synthesizer, the one who put the new ideas together and became their best salesman. It was risky business in the 1960s; those were the years when it was still career-threatening to preach that continents had drifted.

The realization that gigantic plates of the earth's crust actually move in relation to each other was comparable to fitting a key piece into a jigsaw puzzle that suddenly showed how the whole picture goes together. It provided a global solution to geologic puzzles that previously had seemed unrelated; it was a genuine scientific revolution. The Pacific Northwest's jumbled geology could finally be explained by the theory of plate tectonics.

Wilson, speaking at the University of Washington in 1981, said, "I'm giving the same talk now I gave in 1965. The only difference is now you believe it."

❖❖❖

750 million years ago: a piece of the North American continent breaks off and heads out to sea, making Idaho the new west coast.

3

A Continental Split

The story of how the Pacific Northwest was put together begins in a place that must have looked more like a moonscape than a scene on earth. It was a barren, monotonous land of sluggish streams almost completely empty of living things. The story begins before forms of life other than bacteria and primitive algae had appeared on the earth. The time was at least 1.5 billion years ago and probably somewhat earlier; the place, the interior of a continent that included the distant ancestor of North America. (It is not known where on the planet the continent was at that time, nor whether it was part of a larger land mass.)

The beginning came as the continent started to rift, or split itself apart. Huge sections of the earth's crust began stretching apart, thinning and cracking to create a low place into which molten rock repeatedly erupted. Oceans sometimes have been born this way when continents split and the sea fills the depressions between the pieces being pulled apart. But this was what might be called "false labor." Just as the splitting-apart was well started, it stopped. The same thing happened repeatedly over millions of years, and each time the spreading would stall, producing what geologists call a "failed rift."

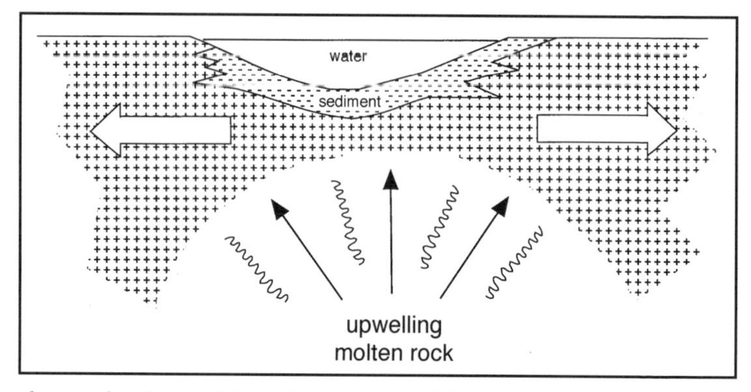

Plumes of molten rock force their way toward the surface, stretching and thinning the overlying continental crust and creating a depression.

A Modern Example

The Pacific Northwest's attempts to split apart about 1.5 billion years ago must have been similar to what is happening in East Africa today. The Great Rift Valley is a much more recent rift, dating back only a few million years to a time when the ancestors of modern humans occupied the area. Earth movement along the rift has uncovered many artifacts and remains, making it a rich area for archaeologists.

Much like the rifting in the Pacific Northwest more than a billion years ago, the movement in the Great Rift Valley appears to have stalled; the trough created by the pulling-apart of the continent is now filling with sediment eroded off higher ground. But a northern extension of the East African rift is still active, and its trough has continued to deepen. The valley floor eventually lowered to below sea level, which allowed the ocean to flood in and form the Red Sea.

The repeated attempts at splitting created a depression—a wide valley—running generally north-south through the part of the ancient continent that is now the Pacific Northwest. What happened next could be expected: Streams running off the land on both sides deposited sediment in the valley over millions of years. Each rifting episode deepened the trough, permitting more sediment to pile into the basin until the sheer weight of accumulating sediment pushed the trough even deeper.

The Belt Rocks

The sediment was laid down where today's Northern Rocky Mountains stand, in parts of Alberta and

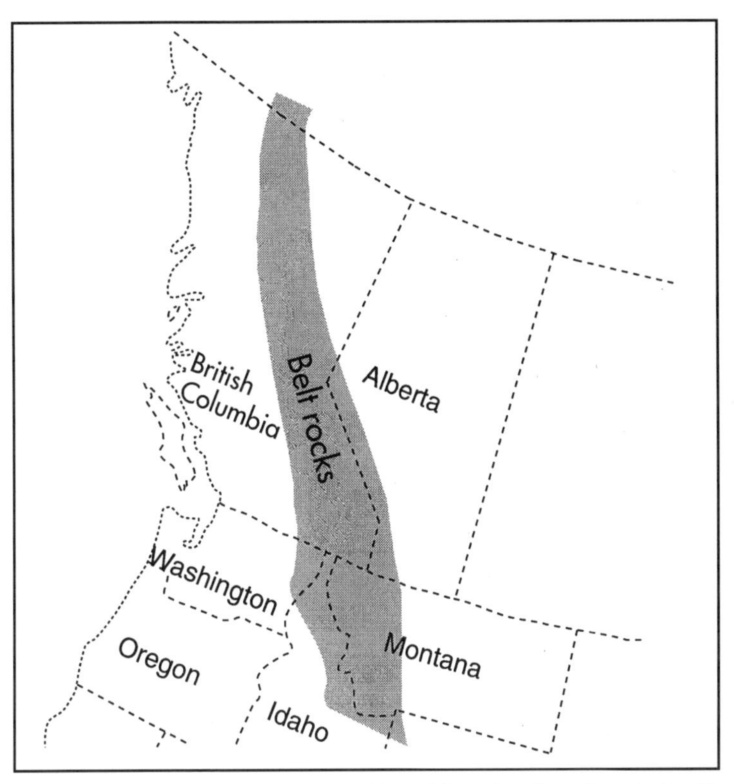

Shaded area shows the approximate position of the Belt rocks, a deposit of sedimentary rock as much as 13 miles thick that accumulated as the continental crust stretched and thinned over millions of years.

British Columbia in Canada and in the western states of Washington, Idaho, and Montana. Researchers estimate the deposits are as much as 70,000 feet thick, something like 13 miles! This makes it one of the largest sediment deposits on the planet. Geologists call the huge deposit, long since pressed into rock, the Belt supergroup for the Belt Mountains in Montana where it was first studied. In Canada, the same rock formation is known as the Purcell supergroup. Here, for simplicity's sake, we will call it (as geologists often do) the Belt rocks.

A fire lookout on 6,298-foot Sundance Mountain stands on rock formed beneath a shallow sea and then uplifted as part of today's Selkirk Mountains in northern Idaho.

By any name, this supergroup is important to the Pacific Northwest story because hundreds of millions of years later new convulsions in the earth pushed up the Belt rocks to form the Rocky Mountains. The rocks high in today's Rockies and adjacent ranges once were mud, silt, or sand in the bottom of the ancient north-south trough.

A Lifeless Plain

The North American landscape must have been largely silent during those eons, the stillness broken only by wind blowing over bare rock, waves breaking on the beaches, the movement of sluggish streams, the fall of raindrops or hailstones, or an occasional thunderstorm. There were no trees or grasses to rustle in the wind, no animals to thump along the ground, no birds to sing.

James W. Whipple, now retired from the U.S. Geological Survey's Spokane office, says the part of the continent bordering the sinking trough, at least on the east side, appears to have been a broad, featureless plain—no mountains, no canyons, no waterfalls. The clues: The sediment deposited in the trough was uniformly fine sand and silt. It was carried by sluggish, meandering streams that branched and reunited in a tangled, braided pattern and finally dropped the sediment onto deltas in shallow water. The Belt rocks contain little of the coarse material that would indicate swift, tumbling streams draining mountainous terrain.

The northwest end of the trough was open to the sea much of the time so that tides alternately bared and covered mud flats. Sometimes rain would make tiny craters in the mud that would dry and harden and then be covered with another mud layer, preserving tiny pits that eventually became part of the rock. Geologists read the rock almost like a book, including the fossil raindrop craters, shrinkage cracks that formed as mud dried under the sun, and crystal casts of salt and other minerals that were deposited as water evaporated. Whipple believes the climate was warm and dry, similar to today's arid regions.

The absence of plants and animals in that long-ago time is both a hindrance and a help in deciphering the past. If geologists know the particular age of plant or animal fossils, they can deduce that deposits containing these fossils were formed during a specific time. So the absence of fossils makes it more difficult to sort out what happened, and when. Except for bacteria and blue-green algae, life forms were almost completely absent on earth when the Belt rocks were forming.

But there's a good side. With no burrowing creatures to disturb sediments, they hardened into rock exactly as they were formed. Scientists can determine the speed and direction of the wind when sand was deposited, the characteristics of streamflow when sediment was laid down, the height of waves blowing across tide flats—even how much the tides rose and fell. Analysis of the

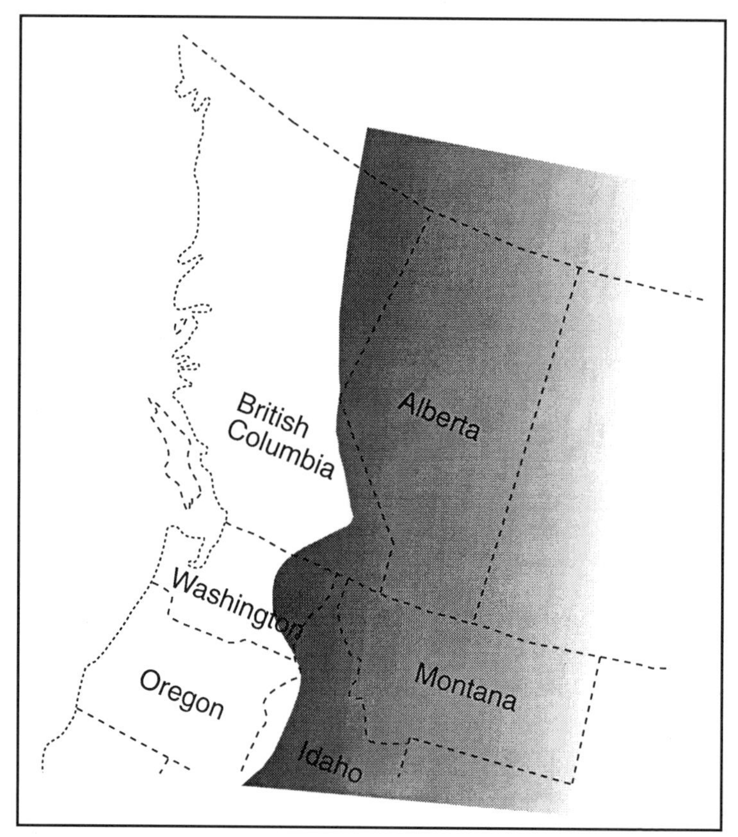

The ancient west coast of North America (shaded area), after a chunk of the continent broke off and drifted west.

minerals located at the bottom of the trough tells whether they formed in the presence of oxygen in shallow water or without oxygen in deep water.

Deciphering the Belt Rocks

Whipple remembers that when he began work as a geologist in the late 1960s in Missoula, Montana, his duties required him to study the Belt rocks that "not many people knew anything about." And, he admitted, "Not many people wanted anything to do with them." The rocks, although brightly colored, were considered mundane. It's possible to drive for hours and still be surrounded by what appears to be the same type of rocks.

Understanding the story in the rocks was difficult because erosion over millions of years had removed complete layers or sets of layers in some places. The Belt rocks, extending for hundreds of miles, have so few distinguishing features that a typical question in the 1960s was whether two patches of rock almost identical in composition were once part of the same formation although now separated by many miles. Whipple says that only in recent years "are we getting to the point where we can say that the unit we see in the Northern Rockies of Canada is absolutely related in time and space to the same unit we see in Montana and adjacent states."

To motorists, the endless Belt rocks may seem monotonous. But, as Whipple says, a trip through them takes drivers "through millions of years of time." The journey through time can be experienced by driving in Glacier National Park in Montana, passing rocks that have been transformed from mud and silt laid down in a soggy slough millions of years ago and which now are part of spectacular mountain peaks and canyons.

The Continent Breaks Apart

Another crucial step in the building of today's Pacific Northwest began about 750 million years ago when, after almost a billion years of failed rifting, the continent actually began to split apart. The split occurred more or less along the north-south trough, the Belt-Purcell basin. As the split developed and the land on one side began drifting westward, the sea filled the gap. The rift, essentially a thin place in the earth's crust, allowed molten rock to erupt under the water, forming new ocean floor beneath the widening gap separating the parts of continent. (The same thing is happening today in the Red Sea.)

The big split not only sent a sizable chunk of continent—no one really knows how big—sailing off to the west, but also established a new ancestral west coast for North America. That ancient coastline ran generally through eastern British Columbia, cut across the eastern edge of Washington, and then south through western Idaho.

As the gap between the continent and its departing cousin grew wider, molten rock continued to erupt underwater and form new ocean bottom. But for the next several hundred million years, the new coastline was a geologically quiet area far from the jostling edges of crustal plates.

Where did the broken-off piece of continent go?

The departing fragment of the continent would have taken thousands or millions of years to disappear over the horizon. Various

---

The Original North America

In the eons that continents have been drifting across the surface of the earth, it was inevitable that, every now and then, several continents and continental fragments would collide and stick together. The foundation of the North American continent was formed by this patchwork assembly process about 1.5 billion years ago. Geologists refer to this original core of rock as the continental shield. This massive body of ancient rock extends east of the Rocky Mountains, underlying much of the central United States and Canada.

Indeed, the term "Canadian shield" is often used because most of the core's exposed rocks are in Canada, particularly in Quebec and Ontario. In the United States, the core is largely covered by sedimentary rocks formed when ancient seas covered much of the central part of the continent.

Fragments of other continents later joined the continental shield to form the present-day coastal areas of North America. The lines where ancient continents met can still be distinguished because rocks on either side of the boundary may differ in age, chemical composition, and geologic environment. Also, when continents collide, the impact invariably pushes up a mountain range. Mountains dating from past collisions have mostly eroded away, but their roots still indicate ancient continental boundaries.

speculations based on similarity of rock types and ages suggest that it is now part of Siberia, or perhaps Australia, or even Antarctica. Undoubtedly, it is still somewhere on the surface of the earth. Continental material is lighter than oceanic crust and therefore can't sink into the earth. Erosion can't reduce it lower than sea level.

The Windermere Supergroup

Sand, mud, and silt continued to erode off the continent into a shallow sea off the new coastline. The accumulating sediment, resting atop Belt rocks, pushed the formation even deeper. This younger layer, now rock, is known as the Windermere supergroup. Largely eroded, it is found mostly in Canada, only rarely in Washington.

In uplifted places today, geologists can find the top of the Belt rocks with the younger Windermere deposits lying on top. The transition from Belt to Windermere rocks records the final rifting of the old continent even though, to an untrained eye, it all looks alike.

The western edge of the continent after the break-up would remain quiet for millions of years until events on the other side of the planet changed North America's west coast into one of the most geologically active regions in the world.

❖❖❖

175 million years ago: on the other side of the planet, the earth's crust splits, generating forces that begin assembly of the Pacific Northwest.

4

Head Bone Connected to the Toe Bone

Remember the old song that runs something like this: Toe bone connected to the foot bone, foot bone connected to the ankle bone, ankle bone connected to the leg bone...and so on up to the head bone. The idea that whatever happens to the toe bone eventually affects the head bone is one that could apply to the chain of geological events that shaped the Pacific Northwest. Millions of years ago, on the other side of the planet, something happened that had an enormous and continuing effect on our part of the country.

A Supercontinent Breaks Up

Almost all the planet's dry land had gathered into one supercontinent, part of a process that apparently has been going on since the earth formed. As continents ride piggyback on creeping ocean plates, they collide and jam together to form a larger continent. (Later, changing forces may break up the composite continent into fragments that go their separate ways.) About 200 million years ago, the continents familiar to us today—North and South

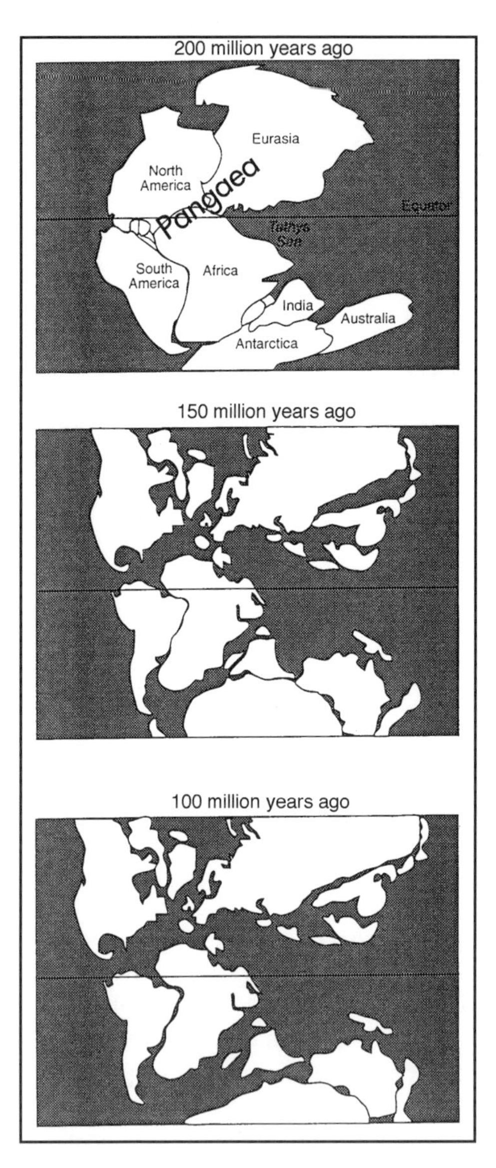

The earth's continents were clustered together 200 million years ago. But by 150 million years ago, the eastern and western halves of the supercontinent of Pangaea had begun to split apart, creating a gap that would become the North Atlantic Ocean. By 100 million years ago, Australia and India had begun their northward journey.

America, Africa, Europe, Asia—were all crowded into a supercontinent called Pangaea. Africa was nestled against North and South America. Europe and North America snuggled with the future Greenland between them. Except for scattered fragments of continental material, the rest of the earth's surface was covered by a vast ocean.

About 175 million years ago, give or take a few million, the supercontinent began to split for reasons that are not completely understood. The process was similar to the rifting that much earlier—before Pangaea had come together—had split off the western part of the continental fragment that was to become North America. But this was much more extensive because it ran right down the middle of all the earth's clustered continents.

To begin with, a depression would have appeared that streams began filling with sediment. At first, the valley must have been narrow enough to see across if anyone had been there. But as the continent continued to stretch apart, the depression got wider and deeper. Eventually the streams brought enough water into the valley to create a lake in the low spot. And finally the bedrock of the valley was lowered enough that the ocean found a way in, forming a narrow inland seaway.

The Mid-Atlantic Spreading Center

Geologists believe the mid-continent separation allowed magma to rise toward the surface, further spreading and thinning the crust. At some point, the molten material pushed up into the valley, perhaps after the sea had flooded it. As the lava cooled, it formed new ocean bottom in the widening trench.

As molten rock continued to push up into the floor of the seaway, it built a ridge that eventually grew into an underwater mountain range. Gravity pulling at the slopes of those submerged mountains caused layers of rock to slide off the mountainsides. Because the mountain range ran more or less north-south, one of these massive slabs slid eastward, carrying the future Europe with it. The slab on the other side of the ridge slid west, with North America as a passenger. As the breakup of Pangaea continued and the Atlantic Ocean grew wider, the continents assumed the shapes familiar to us today.

The ancient spreading center in the Atlantic is still at work, with newly formed sea floor sliding laterally off the ridge. The Atlantic is widening, with the distance between the continents increasing ever so slightly year by year. In fact, Paris is now about

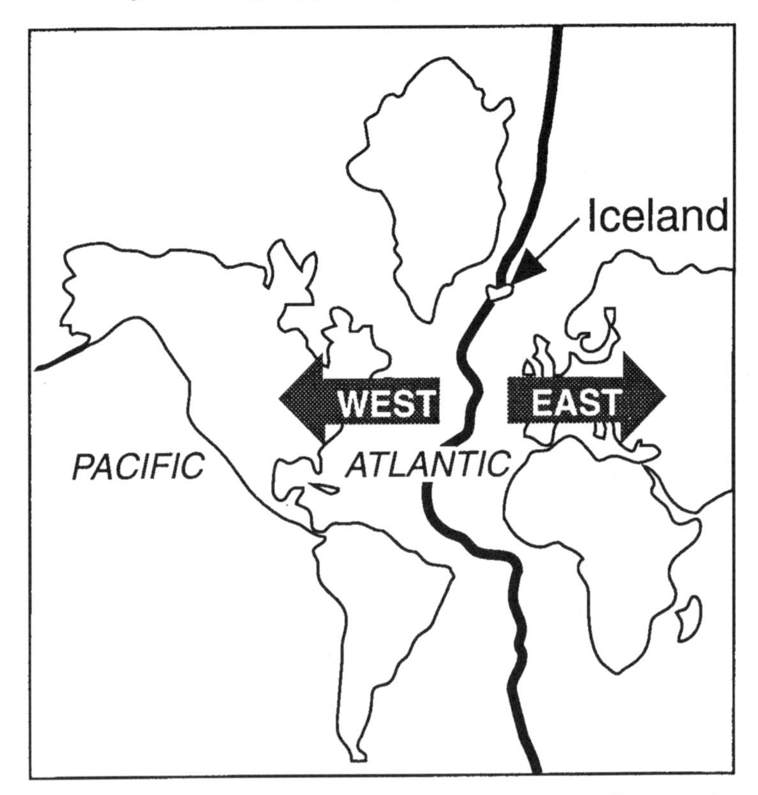

North America and Europe continue to move away from a spreading center in the mid-Atlantic Ocean. Iceland, an area of intense volcanic activity, straddles the spreading center.

50 inches farther from New York than it was in 1927 when Charles A. Lindbergh flew across the Atlantic.

On a Collision Course

Pressure from the mid-Atlantic spreading center changed the character of North America's western edge. For hundreds of millions of years, the coastline had been passive, far from the action of clashing plates. With the opening of the Atlantic Ocean, a segment of the earth's crust began creeping west, carrying all of North America with it. The North American plate, as the huge land mass is known, includes the continental United States, Alaska, Canada, and Mexico, and the western half of the Atlantic floor.

The westward movement of the huge North American plate set the stage for a monumental collision. For millions of years, the Pacific Ocean plate had been creeping east toward the supercontinent. Now, with the North American plate moving west, these two gigantic plates were on a collision course. Obviously, something had to give, and the heavier ocean-bottom plate began sliding under the western edge of the continent, pushing up the thick layers of accumulated sediment just off the coastline. The meeting of these two land masses was the first of many collisions affecting the Northwest. Eventually, piece by piece, a unique and varied landscape was assembled from fragments of once-distant islands and continents. And all this was the end result of a process set in motion millions of years ago and thousands of miles away: a clear case of the head bone being connected to the toe bone.

❖❖❖

150 million years ago: North America collides with and annexes a minicontinent.

5
Wandering Minicontinents

It started when the ancient western edge of North America began slowly moving west, pushed from behind by a widening rupture in the earth's crust that was creating the Atlantic Ocean thousands of miles to the east. This was a significant change for a coastline that had been quiet, where nothing much had happened for hundreds of millions of years. Now it was headed west toward a disparate collection of reefs and islands, submerged mountains, and other chunks of continental material moving east from the Pacific Ocean.

The fragments were riding aboard a vast segment of ocean bottom being pushed toward North America and away from a spreading center where molten rock was creating an undersea ridge. Visualize a series of mountain ranges—some submerged, some rising above the surface—carried on a conveyor belt through the deep ocean; the conveyor belt is the ocean bottom, a rigid platform of basalt sliding toward North America.

The situation promised a violent collision and that's what happened, but in extremely slow motion. As the distance between the continent and the islands narrowed, some of the island groups began crowding and colliding with each other. It wasn't neat and tidy. As odd-shaped segments came together offshore, their borders

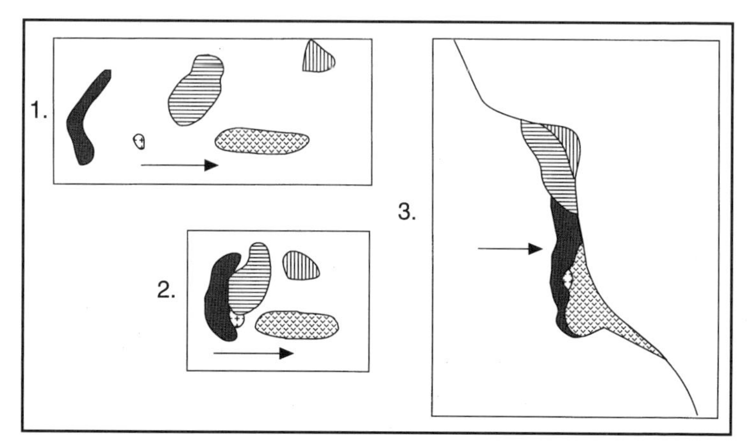

Scattered islands, underwater volcanoes, reefs, and fragments of continents in the ancient Pacific Ocean jammed together as the ocean plate carrying them collided with North America.

were mashed to fit. Waterways separating them narrowed, became cut-off basins, and finally were squeezed and filled out of existence. Rocks crumbled under unremitting pressure. A segment of rock caught between two bigger ones might be pinched into a different shape, or even ride up on top of its new neighbor. Sometimes volcanoes erupted on the bigger pieces. But it all happened very, very slowly over millions and millions of years.

The slow, violent process was an important step in assembling British Columbia west of the Rocky Mountains, Washington except for the eastern edge, and all of Oregon—odd pieces that drifted here from someplace else. But the spot on the planet that now holds the Pacific Northwest was still open sea.

A Wandering Minicontinent

As the batch of islands approached and collided with North America's western edge, the biggest collection of all was approaching: a jammed-together combination of oceanic islands, reefs, seamounts, active volcanoes, and the floors of shallow seas. At first the huge conglomeration of islands—actually a wandering minicontinent—would have appeared against the western horizon, perhaps with volcanoes erupting. As the creeping ocean plate forced the minicontinent toward the coast an inch or two a year, the jumbled land mass began touching the continent, first with its projecting headlands crunching into the coastline. As it advanced, the coastal sea diminished to isolated basins, and then the basins vanished. Layers of sedimentary rock that long before had been deposited in orderly layers offshore were bulldozed, telescoped, folded like an accordion, or even detached in giant slabs and shoved inland. The compression triggered volcanoes along the collision zone.

The ocean-bottom conveyor belt had been plowing into and diving beneath lighter, softer sedimentary rocks underlying a shallow sea off the coastline for millions of years even before the minicontinent arrived. It was an ancient subduction zone, similar to the one we have today where the Juan de Fuca plate subducts beneath the Pacific Northwest.

The minicontinent's cataclysmic arrival about 150 million years ago marked the first great addition to North America since the continent lost its western edge hundreds of millions of years earlier.

A New Coastline

The western edge of the North American continent was then east of today's coastline so the collision zone between continents was about where Idaho, Oregon, and Washington meet today. And even before the minicontinent crunched against that ancient coastline, other groups of islands had arrived, including the building blocks of today's Blue and Wallowa Mountains. They wouldn't have looked much like today's beautiful mountains of northeastern Oregon and southeastern Washington. They were the wreckage of bare volcanic islands, still partially submerged, jammed against North America.

Meanwhile, to the south, one offshore island cluster after another was being forced against the coastline, each one shoving beneath its predecessor. That's why the foundation of the Klamath Mountains is a series of layers stacked like leaning dominoes. The Klamaths we know are in Northern California and Oregon's southwest corner, but these collisions were on the ancient coastline far to the east. The Klamaths moved west to their present position and were uplifted millions of years later.

Fossils from Far Away

When the minicontinent smashed up against North America about 150 million years ago, it carried, somewhere in the jumbled mess, a group of rocks containing fossils formed thousands of miles away. The story of those rocks, which startled and puzzled modern geologists when they found them in British Columbia, began millions of years before the minicontinent collided with North America. In fact, it was so long before that a vast ocean stretched most of the way around the planet, absent only where the earth's dry land was clustered together in one massive supercontinent. It was even before dinosaurs appeared.

What happened was that molten rock began squeezing through cracks in the ocean bottom not far offshore from land that would become Southeast Asia. Such ocean-bottom eruptions are not uncommon; they occur beneath oceans today, although why the earth opens at certain times and places is not fully understood. These ancient eruptions continued for millions of years, eventually building lava mountains so big that they poked through the ocean surface as islands. As the islands grew, their weight caused the earth's crust to sag, and the islands began to sink. Wave erosion leveled and cut the islands into tropical reefs and shoals; the same thing has happened to the string of older islands in the long

Hawaiian chain that stretches northwest of the newer cluster of islands more familiar to tourists.

As more millions of years passed, microscopic forms of animal life lived and died in the warm, shallow water that covered the sinking islands. They sank to the bottom, and in time their skeletons formed a layer of limestone more than a mile thick. The limestone shoals rode eastward aboard the creeping ocean bottom; and eventually they connected with other islands and reefs in the wandering minicontinent approaching North America's west coast. The only way these particular limestone shoals differed from their jammed-together neighbors was that they had traveled the farthest of all.

The Cache Creek Rocks

The rocks were discovered in British Columbia in the 1950s. They puzzled geologists because they were found high in the mountains and several hundred miles from the sea; yet they contained fossils of marine animals. The out-of-place rocks, named the Cache Creek formation for the place in British Columbia where they were first identified, remained a puzzle for years. Adding to the mystery, the Cache Creek fossils belonged to several ancient species that previously had been found only in Japan, China, Malaysia, and the East Indies.

Finally in the 1970s the developing idea that the earth's crust is composed of gigantic, rigid, moving plates began to provide an explanation. It is now pretty well accepted that the ocean-dwelling animals fossilized in the Cache Creek rocks really had lived and died in the tropical waters of the Pacific—more specifically in the shallow water of the Tethys Sea off the coast of Southeast Asia back when the continents were still joined. Rock embedded with fossilized creatures was carried aboard an ocean-bottom plate—or maybe a series of several plates—to North America about 150 million years ago.

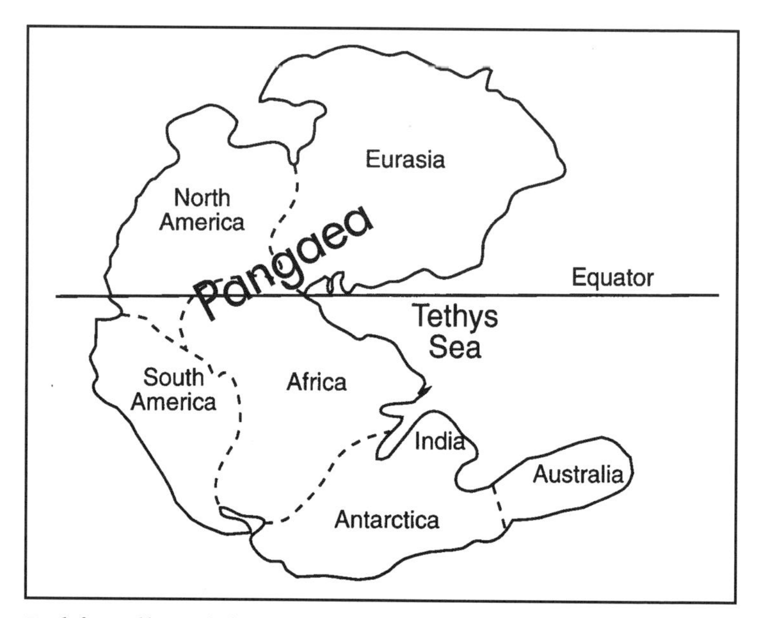

Rock formed beneath the ancient Tethys Sea when the continents were still clustered together is found today in the mountains of British Columbia.

The Cache Creek formation was one of the first to be called a terrane, a word used by geologists to describe a large piece of the earth's crust that has its own geologic character as compared to its neighboring terranes. The boundaries between terranes are faults—places where moving terranes have jammed into one another, or slid under or over another, or slipped sideways past each other. At some of the more obvious terrane boundaries, such as those exposed by erosion, a geologist can walk 100 feet and see a complete change of rock type and age.

Superterrane I

The minicontinent that crunched into North America about 150 million years ago was a massive collection of individual terranes pushed together into one solid land mass. Geologists call it Superterrane I.

Some 200 terranes have been identified along North America's west coast, most of them in Oregon, Washington, British Columbia, and Alaska. Each has its own distinctive history. Almost all have moved from their point of origin, with the Cache Creek terrane coming the farthest of all. Attempting to decipher how the Pacific Northwest was put together is a little like working a jigsaw puzzle backwards: We are confronting the already assembled puzzle and trying to determine where each piece came from and how it got here.

The Cache Creek rocks contain clues to its story. Geologists like to describe the arrival of terranes as though they were ships docking with a continent. During the long journey from its place of origin and sometime before it "docked" with North America, the Cache Creek terrane became part of a larger land mass. Today Cache Creek is squeezed between two bigger terranes, Quesnellia on the east and Stikinia to the west, both named for the regions in British Columbia where they were identified. Neither of the bigger terranes contains evidence of an origin in the tropical Pacific Ocean.

Obviously no one knows exactly what happened as the wandering Cache Creek terrane approached North America. But we can surmise that it encountered a cluster of volcanic islands that were being squeezed together as they, too, inched toward the continent. Using clues left in the rock, geologists believe Cache Creek drove against the larger Quesnellia, itself an assembly of former islands. As Cache Creek collided with Quesnellia, it was shoved from behind by the biggest of all west coast terranes, Stikinia. In the collision, Stikinia dived beneath Cache Creek, causing the smaller terrane to ride up and over. Small wonder that Cache Creek is a jumble of broken rock.

Eventually, Quesnellia, Stikinia, and smaller islands formed Superterrane I, a huge patchwork minicontinent with Cache Creek firmly caught in the middle. The arrival of Superterrane I briefly stopped the subduction of the ocean plate beneath the continent because the arriving minicontinent was simply too bulky to be dragged down with the ocean plate. It stuck against the continent and plugged the subduction zone.

Although Superterrane I caused great disruption as it plastered itself along hundreds of miles of coastline, it could not permanently stop the subduction process as North America and the ocean plate continued irresistibly on a collision course. The creeping ocean plate had to go somewhere so it broke on the ocean side of Superterrane I, creating a new subduction zone and enabling the ocean bottom to continue diving beneath North America.

Superterrane I was big, perhaps about the size of California, although there's no way to guess its shape. So by somewhere around 150 million years ago we have this huge addition stuck to the west coast, while the ocean plate continued to dive beneath the continent along the entire west coast between the latitudes of today's British Columbia and Mexico.

As the ocean plate angled deep beneath the edge of the continent and reached a depth of 40 to 50 miles, it became so hot it began to melt overlying rocks of the continent. This molten material, hotter and lighter than surrounding rock, began crowding its way toward the surface in great pools, pushing aside surrounding rock and lifting the surface into mountain ranges. Some of it reached the surface and erupted as volcanoes but most cooled underground in great pools of granite-like rock called batholiths. The Idaho, Sierra Nevada, and Southern California batholiths began to form about the time of the docking of Superterrane I.

Superterrane II

The relentless eastward creep of the ocean floor was also assembling another mid-ocean superterrane for delivery to North America's west coast. But unlike the Cache Creek rocks in

Batholiths and Plutons

Plutons are bodies of rock that once rose from a great depth as magma, intruded into older rock above, and then cooled, never reaching the surface of the earth. Batholiths are exceptionally large plutons, usually composed of a number of individual flows. The Sierra Nevada, the backbone of California, is the core—the remaining batholith—of what must have been a much bigger mountain range that has been reduced by erosion.

As batholiths cool gradually underground, they serve as molten rock reservoirs that feed nearby volcanoes. Batholiths remain long after the volcanoes have died and eroded away. Unlike volcanoes, which are built of easily eroded layers of crystallized lava and ash, batholiths are relatively resistant to erosion; the magma cooled slowly underground as a single, solid mass.

Where erosion has exposed a pluton, making it accessible, the ancient rock can provide geologists with a window to the past. For example, a line of plutons may mark the zone where a superterrane collided with the continent. The age of rock in the plutons is a clue to when the collision occurred.

Today's Cascade Range south of Snoqualmie Pass in Washington State is underlain by a batholith formed by subduction along today's coastline. In fact, Mount Rainier sits atop a slowly cooling pluton that has fed the volcano in the past and probably will do so again.

Superterrane I, this conglomeration of oceanic islands and bits of ocean bottom did not contain clues of travel from the western Pacific. Instead, it carried fossils indicating it had once resided in waters not far off the coast of the Americas.

This was Superterrane II. By this time, the ocean plate's direction of travel had veered from almost due east to northeast. So Superterrane II probably approached North America from the southwest. In ultra-slow motion it jammed into the continent about 100 million years ago with the same cataclysmic results as that of its predecessor. Sedimentary layers beneath the coastal sea folded like an accordion ahead of the arriving superterrane; slabs of the continental edge detached and were pushed inland.

As the superterrane drove ashore, it forced wedges of ocean bottom under the edge of the continent. As the slices were dragged down to depths of 12 or so miles, high pressure caused chemical changes in the ocean-bottom rock, changes that would provide clues for geologists millions of years later. Eventually, the pieces of altered seafloor were forced to the surface and folded into mountains.

Superterrane II itself, too big to dive beneath the continent, plugged the subduction zone. As happened earlier with the docking of Superterrane I, the disruption forced the ocean plate to break outside of the collision zone, forming a new subduction zone.

Where did these superterranes dock?

Although the question is debated, there is evidence that when Superterrane I jammed against the ancient coastline of North America, it was spread out over hundreds of miles extending from the latitude of today's British Columbia to just south of the Oregon-California border. (Cache Creek and Stikinia make up parts of today's British Columbia and Alaska. Superterrane I's other big terrane, Quesnellia, makes up the Okanogan Highlands in northeastern Washington and southeastern British Columbia.)

Superterrane II also would have jammed against hundreds of miles of coastline. But there is increasing—and controversial—evidence that it docked with the continent at the latitude of today's northern Mexico. Superterrane II's most famous terrane is Wrangellia, named for the Wrangell Mountains in southeastern Alaska where it was identified. (Wrangellia also includes Vancouver Island, the Queen Charlotte Islands, much of British Columbia's coastal strip, and parts of the Yukon and Alaska.) If Wrangellia docked along the Mexican coast, it must have later moved more than 1,000 miles to the north. (If Superterrane I has crept north from the point where it docked, it must not have moved very far, perhaps a few hundred miles.)

Is it possible that these huge chunks of real estate have moved hundreds of miles, and perhaps a thousand or more, from where they first joined North America? This idea poses obvious problems for geologists studying how the Pacific Northwest was put together.

❖❖❖

Study of ancient rocks in the North Cascades suggests that parts of the Pacific Northwest may have come from 1,000 or more miles away.

6

Land on the Move: Baja to British Columbia

The 1970s were perplexing times for traditional geologists who had always wanted to see direct, visual evidence in the ground of what had happened at a certain locality in the past. They were faced with the new idea that much of the Pacific Northwest had formed somewhere else and then was moved here for final assembly. Much of this new theory lacked the familiar kinds of clues that could be dug up, looked at, or studied in the laboratory.

But specialists in paleomagnetism, a new branch of geology, were turning up a different kind of evidence showing that many rocks had been formed more than 1,000 miles south of where they now are located. Paleomagnetism involves the study of ancient magnetic clues in rock. The new information must have been astounding to conventional geologists who had spent their careers figuring out the history of a region, basing their reconstructions on what they could see in the rocks.

A handful of American and Canadian geologists began applying the science of paleomagnetism to Pacific Northwest and Alaskan rocks in the 1970s. Here's how it works: when rock is still molten, its tiny magnetic particles can move freely. They line up like free-swinging compass needles, pointing toward the earth's magnetic poles. When the rock cools and solidifies, those "compass needles" are permanently fixed in the rock. Using paleomagnetic techniques, scientists can read those natural compasses in rock samples millions of years after the rock has cooled.

Ancient "Compasses" and "Clocks" in Rock

Near the equator, the earth's magnetic field causes magnetic particles in molten rock to line up parallel to the earth's surface. But nearer the poles, the particles begin to point toward the center of the planet; at the magnetic poles they will point straight down. Thus these permanent natural compasses carry an approximate record of the latitude at which rock cooled from its molten state.

The ancient "compasses" may carry another clue to the past. If they no longer point to magnetic north, it is assumed the rock has rotated since it was formed; sometimes a series of samples will indicate that a whole mountain range has rotated.

Volcanic rock also contains clues to how much time has passed since it cooled from a molten state. Age can be approximated by measuring the decay of naturally radioactive elements locked into the rock when it cooled, something like reading a radioactive clock that began ticking when the rock crystallized.

In a complicated way, Nature has played into the hands of scientists equipped with modern techniques by making these deeply buried volcanic rocks accessible. A collision between creeping ocean bottom and a continent typically pushes up a mountain range; eventually the mountains erode, exposing the underlying batholith where most of the magma cooled. Where a batholith is exposed by erosion, geologists scrambling around the back country can easily get at it to take samples. With a lot of work and some luck, the samples can reveal the age of a batholith and, thus, the approximate time of a superterrane's arrival as well as the approximate latitude where it docked, and whether the rock has rotated since it solidified.

Myrl E. Beck, Jr., a Western Washington University professor of geology and one of the pioneers in paleomagnetism, turned up evidence in the 1970s that the Mount Stuart batholith in Washington's North Cascades apparently had formed 1,500 or more miles to the south and, furthermore, had rotated clockwise 60 degrees. At first, his suggestion that these findings indicated large-scale movement along the coastline was largely disregarded. He can remember being laughed at by other geologists at professional meetings.

Baja–B.C. or Bust

More research strengthened the idea that Superterrane II, which included the Mount Stuart batholith in Washington and others in British Columbia, had indeed docked with the continent at about the latitude of today's northern Mexico. Superterrane II acquired the nickname of Baja–B.C., indicating it traveled from Mexico to British Columbia. The nickname grew more serious as studies began indicating that Baja–B.C.—oceanic islands and reefs, batholiths, and mountains pushed up and valleys pushed down during the collision—had traveled as a more-or-less intact block for the first part of its journey north along the coast.

The Baja–B.C. concept is still difficult for traditional geologists to accept. For one thing, all geologists want to see evidence in the ground of what happened in the past. And so far no one has found a major, through-going fault that would mark where the inside edge of Baja–B.C. slid along the continent. Believers in Baja–B.C. say that great earth movements and changes since the time of the superterrane move could have hidden the fault so well that, if it's there, it may never be found.

The Stuart Range, Washington State

Tilt Versus Travel

Critics of the idea of Baja–B.C. doubt that much of the Pacific Northwest could have collided with the continent far to the south and moved such a vast distance to the north. According to these geologists, plutons in the ancient Superterrane II could have formed at their present latitude if there was some other factor that could account for the seemingly out-of-place paleomagnetic measurements; perhaps, they speculated, the plutons have tilted considerably since they crystallized. The weakness of this theory is that it would require all plutons from British Columbia to Washington to have tilted the same amount (about 30 degrees) in the same direction (to the southwest). Myrl Beck, the Western Washington University pioneer in paleomagnetism, dismisses the idea of tilt, saying it would have required "a curious act of God."

There's no direct way to determine whether a pluton has tilted because a pluton solidifies into one solid mass without the layers seen in sedimentary rock. However, researchers used a new technique to try to find evidence of tilting at Mount Stuart in the North Cascades. Little evidence of tilting was found, and instead of supporting the tilting idea, the study actually strengthened the theory that the superterrane traveled long-distance from Baja to British Columbia.

In another study of what was once Superterrane II, researchers found layered volcanic and sedimentary rock—still horizontal, and clearly not tilted—holding magnetic evidence that it cooled more than 1,000 miles to the south.

The idea of Baja–B.C. remains controversial. Robert Butler of the University of Arizona, a leading believer in tilt, told a Geological Society of America meeting in Seattle in the mid-1990s that he was still skeptical of Baja–B.C. Noting the accumulating evidence favoring long-distance movement of Superterrane II, he jokingly bemoaned the lack of a witness-protection program for believers in tilt.

How could Baja–B.C. move more than 1,000 miles?

The superterrane may not only have moved north over 1,000 miles, it may even have passed Superterrane I, which apparently was on a several-hundred-mile journey of its own.

The key lies in the Farallon plate, the creeping ocean plate that had pushed the two superterranes against North America in the first place. Approximately 100 million years ago, the entire eastern Pacific Ocean off North America was underlain by this one giant plate of ocean bottom.

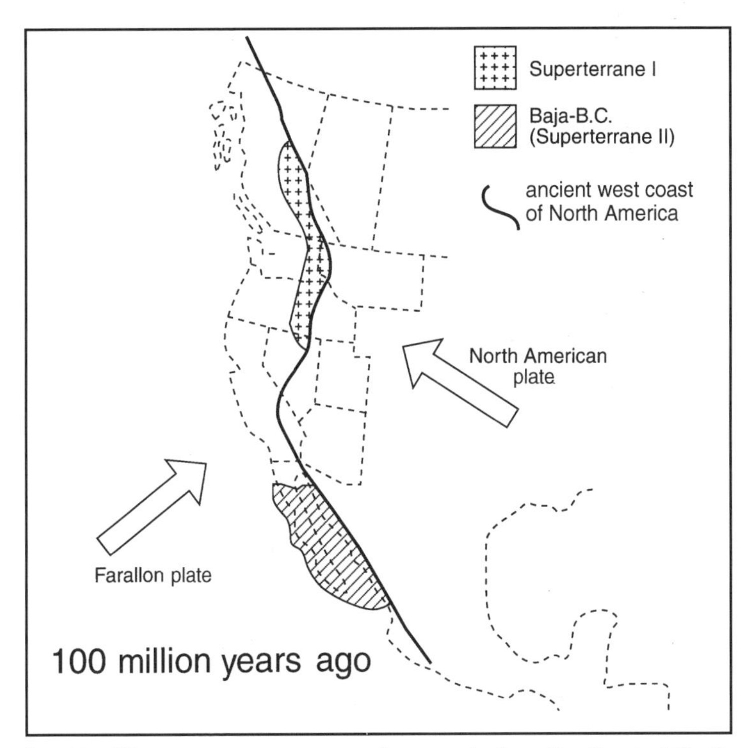

By 100 million years ago, convergence between the huge Farallon and North American plates had squeezed two superterranes (Superterrane I and Baja–B.C., or Superterrane II) against the continent. Their sizes and shapes can only be estimated.

The motion of the Farallon plate, creeping east when it delivered Superterrane I, had shifted toward the northeast by the time Superterrane II arrived. Scientists can reconstruct these movements because as plates move they leave tracks in the ocean bottom—stripes of differently magnetized rock, ridges that once were spreading centers, submerged volcanoes, and bands of shattered rock where plates once rubbed against each other.

Around 100 million years ago the forces that move plates were continuing to change. The Farallon plate veered even more toward the north and its speed increased, to something like 3 or 4 inches a year. And about 84 million years ago, something happened that caused three pieces to break off the northern part of the Farallon plate and begin moving independently. One of them, the Kula plate, figures in our story.

The "Kula Express"

The Farallon plate already was moving northeast when the Kula plate broke off and began heading north. Farallon's northward motion would have added to the velocity of the Kula plate, much as a tail wind increases an airplane's speed. The Kula plate appears to have traveled as much as 6 inches a year, moving 100 miles in a million years. That's pretty fast in geological terms.

The story of the Kula plate is difficult to reconstruct because it has disappeared, subducted beneath the North American plate along today's Aleutian Islands arc. (The word "kula," borrowed from the Athapascan Indian language, means "all gone.") When the Kula disappeared it took with it clues that would have revealed much about its motion. However, ocean plates that bordered the Kula plate during its relatively short life do retain some of the story.

Scientists are reasonably certain that the Kula plate touched North America's west coast but they are not sure where. No matter where contact was made, it could have provided fast northward transport for a terrane jammed against the coastline, perhaps even carrying along fragments of the continent that had become attached to the terrane.

Geological clues indicate Superterranes I and II were pushed together by about 70 million years ago. This means that if Superterrane II (Baja–B.C.) docked at northern Mexico 100 million years ago, it moved north about 1,000 miles over the next 30 million years; eventually it would have arrived in the neighborhood of Superterrane I, which was still spread out along the coastline

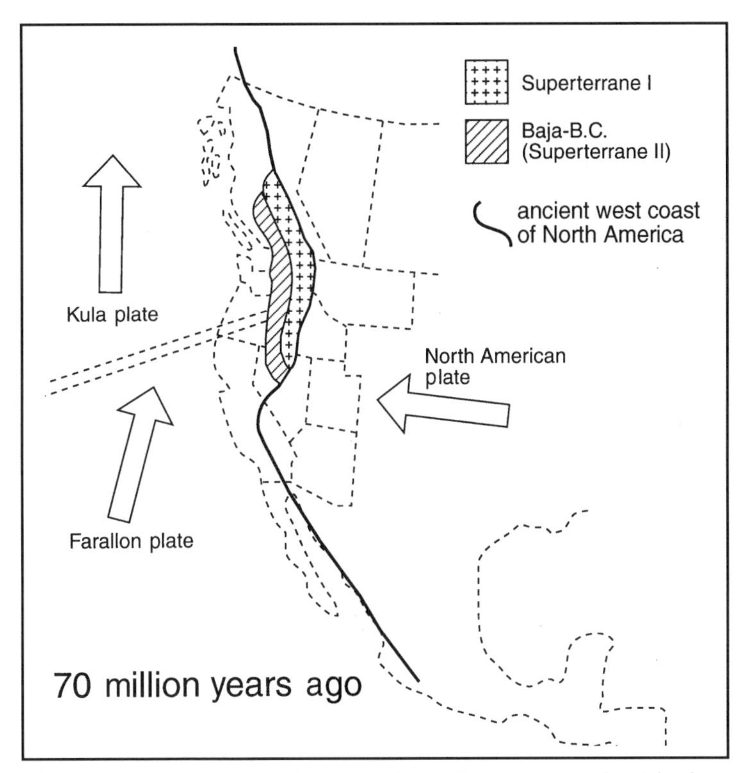

The superterranes had come together by 70 million years ago. The Kula plate, separated from the northern edge of the Farallon plate, was moving north, possibly providing northward transport for Baja–B.C. (Double broken lines indicate the break between the Kula and Farallon plates.)

between the latitudes of today's Oregon-California border and British Columbia.

This is how a growing number of scientists both in the United States and Canada explain this migration: Baja-B.C. began its northward trip about 85 million years ago. It could have been a passenger on the eastern edge of the Kula plate. Or perhaps it was pushed along by the Farallon plate as it slanted obliquely into the continent. Or perhaps it rode first on the fast "Kula Express" and then became detached and was pushed along by the slower but dependable Farallon plate.

At any rate, by about 70 million years ago, Baja-B.C. was colliding with Superterrane I. Then the two superterranes moved north together for another 20 million years or so. During this part of the trip, both Baja-B.C. and Superterrane I began to come apart. The portions on the ocean side were sheared off and moved faster than the eastern part. The result, shown by magnetic clues in the rock, was that western slices of Baja-B.C. traveled farthest, three times as far as those closer to the continent's edge. The most far-traveled fragments today are in southeastern Alaska. Traveling shorter distances were British Columbia's Coast Range, Vancouver Island, and the Queen Charlotte Islands. The once-intact minicontinent had been smeared out along the entire northwest coast.

The fact that sliced-off portions of terranes were sliding past each other may provide another explanation for rock rotation. If a piece of a terrane got caught between neighbors moving at different speeds, it could have been forced to rotate. Geologists refer to this possibility as ball-bearing rotation.

The Superterranes Move On

Superterrane I underwent the same dismemberment as Baja-B.C. but to a lesser degree. A major strike-slip fault, one where terranes slide past each other, apparently opened up in the eastern portion

Sackit Canyon

Exploring the Boundary

Sackit Canyon near the Canadian border in northeastern Washington is a place where ice age glaciers did geologists a favor by bulldozing away the cover of younger rocks hiding the boundary between ancient North America and Superterrane I.

"Almost everyone agrees that east of Sackit Canyon is ancient North America," said Robert Powell of the U.S. Geological Survey in Spokane. West of the canyon are the Okanogan Highlands—the rugged remains of Quesnellia, a part of Superterrane I. Powell frequently visits Sackit Canyon in search of rocks that could tell more about Superterrane I's collision with the North American continent. Unfortunately, the glacier that scooped out the canyon left behind a deposit of its own—glacial debris.

"In Sackit Canyon, you're that close," Powell said, sitting in his small office. "Over the width of this room is a glacial deposit that has beneath it the actual contact." He continued, "Out there I go back and forth between, 'Man, it's going to work, it's going to work, I'm going to learn something here,' to 'No, I can't see through this junk.'"

Powell, who considers himself a devil's advocate regarding fancy theories about terranes, still finds his investigation exciting, hoping that one day Sackit Canyon will yield answers to some of geologists' long-held questions.

of Superterrane I. Traces of faults active while most of Superterrane I moved hundreds of miles have been identified in British Columbia, but they have not been found farther south. Perhaps they have been hidden by the great floods of lava that covered older formations in eastern Washington and Oregon in a much more recent period. By 50 million years ago, both superterranes had arrived. The northern edge of Washington and most of British Columbia were in place.

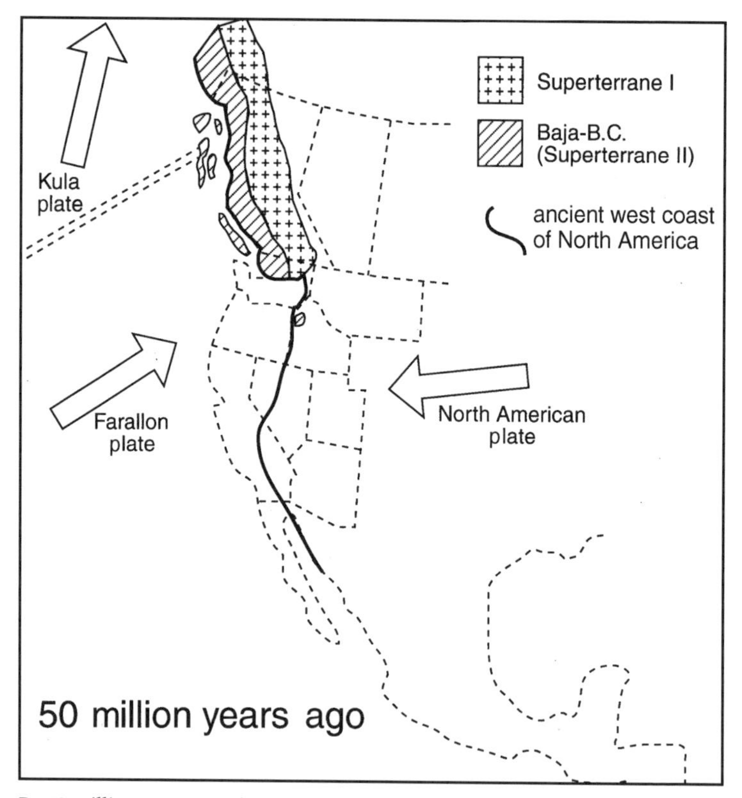

By 50 million years ago, the two superterranes had arrived in the Pacific Northwest. Most of British Columbia and the northern part of Washington were now set in place.

The Rocky Mountains, 90 to 40 Million Years Ago

Alberta, Canada

Processes assembling the Pacific Northwest contributed to building the northern Rockies that run from Canada south to Montana and Idaho. As the superterranes were docking, the Farallon plate continued to subduct beneath North America's western edge, located at that time in eastern British Columbia and Idaho. As the huge ocean plate was pushed down into the trench, most of the lighter sediment covering the hard-rock plate scraped off and piled up on the continent. Other sediment was jammed under the coastline, further raising the continent's edge. With the plate's irresistible advance, miles-thick slabs of continent broke loose and were bulldozed inland. This combination of events raised and thickened a strip of coastline—the beginning of the Rocky Mountains.

As the diving plate reached a depth of 40 or 50 miles, it sent great pools of magma pushing toward the surface. These batholiths cooled 5 or 10 miles below the surface, further raising the continent's edge. Thus around 90 million years ago, there must have been a range of mountains rising from the ocean along the coast. To the east, a vast, shallow sea flooded the continent's interior repeatedly for millions of years.

About 80 million years ago, the North American plate speeded its westward rate of creeping and pivoted slightly counterclockwise, shoving the ocean plate even faster under the continent. Perhaps, the diving slab took on a shallower angle, further wedging up the entire west coast. Even the continent's level interior was pushed up toward the rising Rockies, expelling much of the interior sea. Today, the Great Plains west of the Mississippi River rise like a ramp to the Rockies.

Today in the northern Rockies, there is evidence that thick slabs of continental edge broke loose, and were stacked and bulldozed 50 or 100 miles east by the Farallon plate. Also, slabs of sedimentary rock appear to have been pushed east off the rising granitic batholiths, adding to the future Rockies. For example, Central Idaho's batholith—250 miles long and up to 100 miles wide—now is exposed, making up several mountain ranges. The Rockies were in place by 40 million years ago. Their final carving and sharpening by glaciers began with the ice age about 2 million years ago.

Even though most of Superterrane I slid north, part of the original collision zone is preserved in northeast Washington and eastern British Columbia. This is where the Quesnellia terrane at the leading edge of Superterrane I plowed into and under sediments beneath the shallow sea bordering the continent; it is the actual boundary between material eroded from the North American continent and the huge ocean island that arrived 150 million years ago. The problem for curious geologists is that, except for a few places gouged out by glaciers, the collision zone is covered by younger volcanic rock.

By about 50 million years ago, both superterranes had arrived at about their present locations, dismembered and rearranged since they collided with the continent. There is still controversy over whether parts of the superterranes traveled as much as 2,000 miles or merely several hundred miles. But either way, by 50 million years ago, parts of the Pacific Northwest were assuming a more familiar shape.

❖❖❖

Part II

The Modern Northwest

The Landscape as We Know It and How It Got That Way

Pieces of the continent break off as faults and fractures on the western edge of North America yield to pressure from a fast-moving ocean plate.

7

Strike-Slip Faults and Leaky Fractures

Joseph A. Vance, a University of Washington graduate student in the early 1950s, spent parts of five summers scrambling through canyons and over mountains in the remote western reaches of Washington's North Cascades. The country was wild and still almost unexplored by geologists. Foot travel exacted a steep price in time, sweat, and worn-out knees.

Vance's assignment: 200 square miles of wilderness. Map it. Describe the rocks and where they lie, whether at the bottoms of canyons or atop peaks. Try to visualize the forces that pushed this landscape together.

"Much time was spent merely in brush fighting and backpacking," he noted in his doctoral dissertation in 1957.

But Vance's dissertation also described a puzzling, previously unknown feature he'd found in his square of wilderness: an ancient rift in the mountains that separated rocks of different types and histories. It is unusual for the types of rocks to change completely within a distance of a hundred feet or so. But here they did.

Vance didn't realize it at the time—neither did anyone else—but he had discovered a scar left in the earth from when major parts of the Pacific Northwest were pushing and sliding into place.

"I realized it was a big fault, but I had no idea what a major structure it was," Vance recalled many years later when he was a geology professor at the University of Washington.

The fault, which runs almost due north-south, was created when the western edge of North America was being wrenched apart, twisted by seafloor movement off the coast. One piece of the continent slid horizontally past another along the fault line.

The grinding movement left a band of broken rocks that quickly eroded and was scooped out by streams flowing along the fault line. One of these streams later caught the eye of explorers because, unlike most mountain streams, it ran in a straight north-south line for several miles. They named it Straight Creek. (As it turned out, the fault extends into British Columbia, where it is known as the Fraser River fault. The Fraser River itself flows almost due south along the trace of the fault until the river abruptly turns west at Hope to head toward Vancouver.)

No one paid much attention to Vance's discovery in the 1950s. But revolutionary new ideas swept geology in the next decade or so. The rift, by then known as the Straight Creek fault, began to look very important to the story of how the Pacific Northwest was assembled.

Although the Straight Creek fault is the best documented and most famous in the Pacific Northwest, it is just one of a series of faults that opened as western North America strained under unimaginable forces and began to break apart about 50 million years ago.

The Kula Plate Moves In

The segment of ocean bottom known as the Kula plate is presumed to have been off our coastline during that time. The Kula plate's

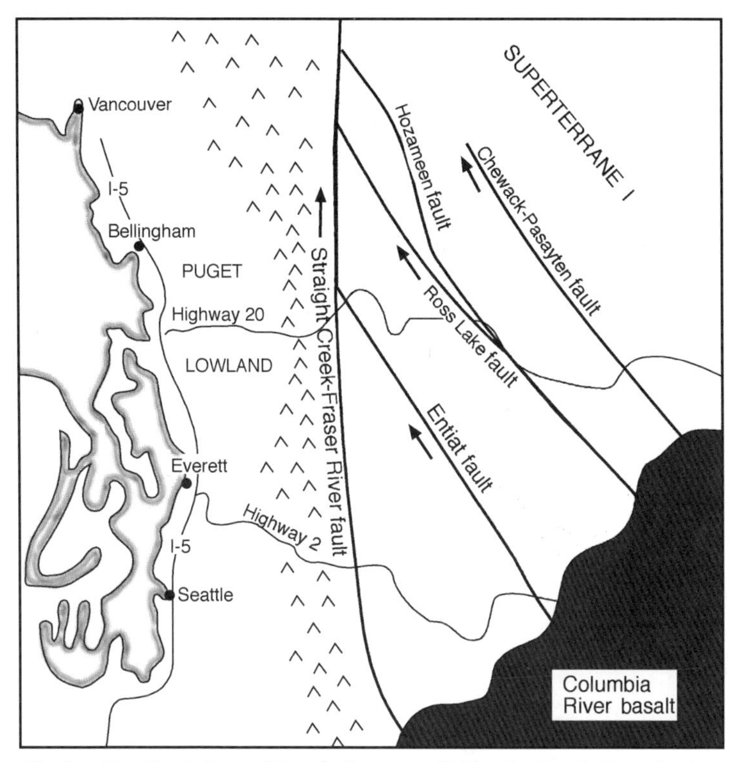

The Straight Creek-Fraser River fault system divides the North Cascades into distinct segments. Arrows indicate ancient movement west of faults.

sharply oblique push toward the north-northeast produced an enormous strain on the edge of the continent until finally the continent broke, allowing movement along faults such as the one later occupied by Straight Creek.

Imagine the western North American continent as a sheet of stiff cardboard so battered that creases, or weak places, are visible. If you apply vise-like pressure head-on against the edge, the sheet eventually will buckle into ridges along the weak places. But if the pressure is applied at a slanting or oblique angle all along the edge of the sheet, the cardboard may twist until pieces tear off along the weakened creases and begin moving. In simplified terms, that is what happened to the part of North America that was to become the Pacific Northwest.

Terranes had been cracked and fractured millions of years earlier as they were jammed against the ancient continent. Under the oblique thrust of the Kula plate, the old fractures crumbled, allowing chunks of the continent's edge to break off and begin moving north. As they moved, these broken-off sections scraped against the continent along the Straight Creek and other faults that had begun slipping in response to the strain.

Strike-Slip Faults

The faults created by the movement of the Kula plate were what geologists call strike-slip faults: terranes on opposite sides of the fault had slid horizontally past each other. They were right-lateral faults, which means that if an observer were facing the fault during an earthquake, the land on the other side would appear to jump to the right. At these ancient faults, the land to the west of the fault crept north as ocean plates pressed against the continent.

The famous San Andreas fault in California is a modern example of a right-lateral strike-slip fault. West of the fault, on the ocean side, a strip of California creeps northwest. Geologists can see where the San Andreas has stepped inland in the past to capture chunks of continent and begin moving them north. The San Andreas has a system of subsidiary faults that distribute the strain, similar to the ancient Straight Creek system of faults in the Pacific Northwest. The Ross Lake fault, a subsidiary of the Straight Creek fault system, eventually became the route taken by the Skagit River. The river is now dammed and the reservoir extends northwest into British Columbia along the trace of the Ross Lake fault.

Leaky Fractures

Meanwhile, the fast-moving Kula plate was carving giant slices off the coastline and carrying them north; some are now believed to be part of modern Alaska. The wrenching of the edge of the continent caused by the Kula's movement pulled open faults in the ocean bottom just off the coast. Geologists refer to these rifts as leaky fractures because, as weak spots in the crust, they provided pathways for molten rock to rise.

The opening of leaky fractures offshore set the stage for another major development in the construction of the Pacific Northwest. Vast quantities of lava (similar to the basaltic lava that is still building the Hawaiian Islands) poured out of these fractures, covering the ocean bottom offshore. This series of mostly underwater eruptions began about 60 million years ago and continued for millions of years; in sheer volume, it was one of the earth's greatest outpourings ever.

The Coastal Mountains: Native Northwesterners

The basalt that solidified from these massive outpourings forms the foundation of today's Coast Range, west of the Willamette Valley in Oregon, and the Willapa Hills, the Black Hills, and the Olympic Mountains in Washington. The north-tugging action of the Kula plate along the coastline may have pulled this mass of basalt northward a little bit. But paleomagnetic measurements show that any movement was small. Unlike the superterranes that had arrived much earlier, the hills and mountains of the Coast Range are native Northwesterners.

The coastline still angled southeast from the south tip of Vancouver Island to western Idaho, and the piles of basalt, mostly still submerged, paralleled that ancient coastline. These offshore

The Hot Spot Theory

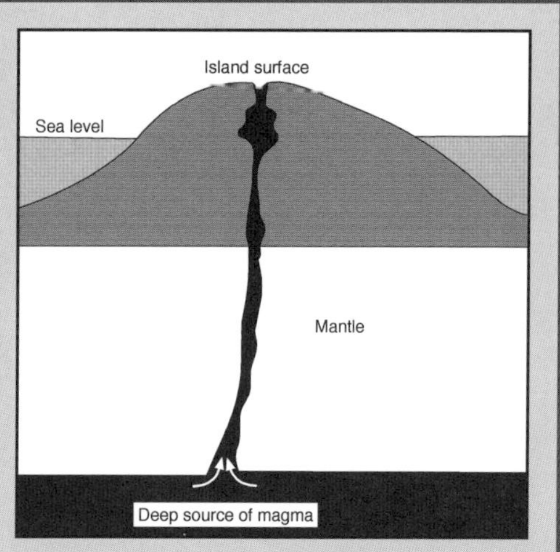

There's another explanation for the origin of the great mass of basalts that form the foundation of the Coast Range. Some geologists and oceanographers believe that a hot spot—a place where molten rock pushes through the earth's crust—was responsible for the eruptions that produced most of the Coast Range basalts.

This theory is based on evidence that there was a hot spot off the coast at the present latitude of southern Oregon approximately 50 to 60 million years ago. It either straddled or was near the spreading ridge between the Kula and Farallon plates. As the hot spot's eruptions formed undersea mountains (some of which reached the surface as islands, others, called seamounts, that did not), movement of the plates carried them away from the spreading ridge. Those volcanic mountains aboard the Kula Plate would have moved generally north; those on the Farallon would have been carried toward the coastline. (The same effect can be seen today on seafloor charts of the South Atlantic Ocean where a hot spot straddles a spreading ridge: One line of seamounts rides a plate west toward South America and another string extends east toward Africa.)

A hot-spot origin would explain why the oldest Coast Range basalts are at the north and south ends, with the youngest in the middle near the Columbia River. And a hot spot as a source would also help account for the huge volumes of basalt. This explanation would change the story of the Coast Range only slightly. The basalts that form the mountains' foundation still would have erupted not far offshore and lined up along the coastline.

More than 100 current hot spots have been identified around the world. Hot spots are relatively fixed in contrast to moving plates, making them useful markers, or reference points, to measure the movements of plates. The source of magma that was once off the Pacific Northwest coast is now known as the Yellowstone hot spot. Over the past 60 million years, North America's westward movement has carried the edge of the continent over the hot spot until today it is beneath Yellowstone National Park in Wyoming.

accumulations of volcanic rock were quickly covered by sediment carried to the sea by ancient streams and rivers.

The Kula Plate Disappears

The Kula plate, which had been wrenching the edge of the continent for millions of years, eventually moved beyond the Pacific Northwest and disappeared beneath Alaska. It was replaced, about 40-50 million years ago, by the Farallon plate advancing slowly from the southwest. As it collided with the edge of the continent, the Farallon plate was thrust beneath it. The subduction zone formed on the ocean side of the massive stacks of basalt on the sea floor, pushing them closer to the coastline.

As the twisting and wrenching of the edge of the continent was replaced by head-on pressure, the Straight Creek fault and its subsidiary faults stopped moving. Batholiths that rose later along fractures associated with the Straight Creek fault system remain unbroken, indicating the fault has not moved since about 35 million years ago.

A motorist driving along today's North Cascades Highway passes through three distinct sections where the origin and type of rocks are completely different. As Vance himself noted, the rock may change very abruptly—within just a few feet—as the motorist crosses an ancient fault. These different segments were once part of separate blocks some distance apart that slipped past each other along the old faults, eventually coming to rest next to each other when the Straight Creek fault system stopped moving.

The Pacific Plate Changes Course

The huge Pacific plate underlying much of the Pacific Ocean abruptly changed course about 43 million years ago. After moving almost due north for millions of years, it shifted toward the northwest. Far across the Pacific, subduction zones opened along

Asia's eastern coastline, pulling the Pacific plate toward them. In response, the spreading center between the Pacific and Farallon plates off the Pacific Northwest swung around, changing the stresses on the edge of North America.

No one knows why the Pacific plate changed course, but there is interesting speculation. One idea connects the change with the collision of the Indian subcontinent against the south coast of Asia. India had been riding an ocean plate north for millions of years and by 43 million years ago it had collided with Asia. The docking Indian subcontinent was too bulky to be pulled down into the subduction zone, and it stuck against Asia. The ocean plate continued to subduct and to push India against the continent, causing compression that pushed up the Himalayas, the world's highest mountains. The pressure from the plate and the rise of the Himalayas continue today. Some speculate that India's collision with Asia may have changed stresses over thousands of miles, even opening up new subduction zones in the western Pacific Ocean.

Whatever the cause, it resulted in the almost unimaginable feat of causing the Pacific plate —a miles-thick block of rock underlying most of the Pacific Ocean—to swerve abruptly to the northwest. Tracks in the ocean bottom reveal the change was almost instantaneous, at least in geological time.

Why are parts of the Coast Range higher than the rest?

The Coast Range stretches from the hills behind Coos Bay on the southern Oregon coast all the way to the Olympic Mountains in Washington. There is a fascinating question about this extensive mountain range: if the whole Coast Range was born in much the same way and then shoved against the continent, why is one section—the Olympics—higher and more spectacular than the rest?

The likely, although unproven, answer appears to lie in differences along the coastline where the Coast Range basalts crowded against the continent. In the southern part of the range—today's

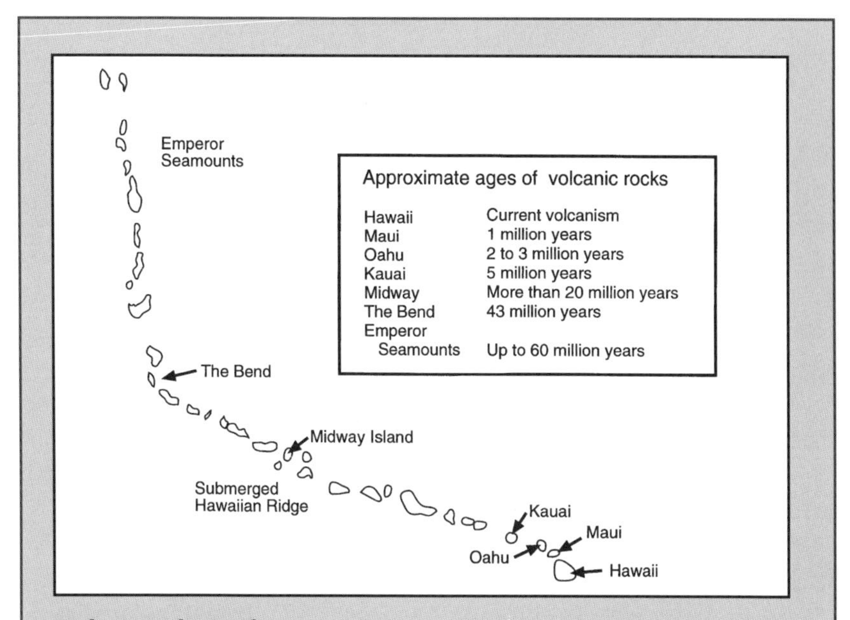

Evidence in the Pacific

The Pacific plate's change of direction 43 million years ago explains an oddity oceanographers have known about for many years: the string of islands, reefs, and sunken volcanoes known as the Hawaiian Islands runs northwest for about 1,500 miles before making an abrupt bend to the north and continuing (mostly beneath the surface of the ocean) for another 2,000 miles.

The islands were built by a hot spot, a fixed plume of molten rock rising through the earth's crust. Over the past 43 million years, the Hawaiian hot spot formed island after island as the Pacific plate crept over it toward the northwest. Volcanic eruptions are still building Hawaii, the big island that still sits over the hot spot.

Radioactive dating reveals that the rocks of the Hawaiian Islands become progressively older to the northwest: Rock on Maui is as old as 1.3 million years, on Oahu 2 million to 3 million years old, and on Kuai more than 5 million years old. The rocks continue to get older, reaching 43 million years old at the point where the string of islands bends to the north. The Emperor chain, as this line of underwater mountains and reefs is called, runs almost due north, eventually ending in remnant formations of volcanic rock more than 60 million years old.

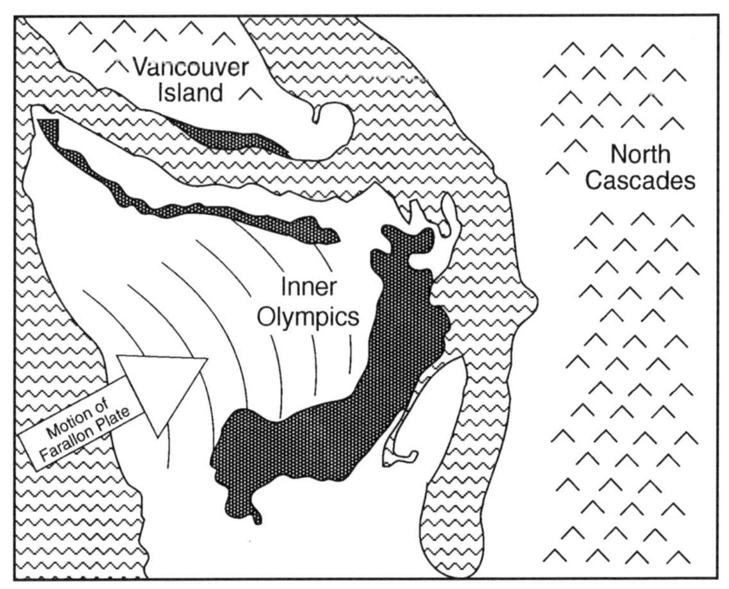

The Crescent formation (shaded area) is composed of layers of basalt turned on edge, which rim the Olympic Mountains on the north and east and continue along the south edge of Vancouver Island.

Oregon and southern Washington—there were no major barriers in the path of the rocks as they crunched ashore, pushed from behind by the Farallon plate.

But the situation was different in the north, site of today's Olympic Mountains. Vancouver Island and the North Cascades, which had arrived millions of years earlier as part of a superterrane, formed a massive corner in the path of the advancing basalt formation with its thick cover of sediment, the future Olympic Mountains. As the former ocean bottom pushed into the immovable corner, something had to give, and it apparently was the arriving formation. The rigid layers of basalt actually tilted on end and bent into a horseshoe shape as they were forced into the corner formed by Vancouver Island and the North Cascades. This explains why a rim of basalt—the layers turned on edge—curves around the inner Olympics on the east and north sides. It's known as the Crescent formation.

The exposed edges of the basalt layers give an idea of the size of the Crescent formation. At one point the edges are 12 miles wide, which means the formation was at least 12 miles thick when it formed on the ocean bottom.

As the basalt tilted, sediment lying on top was folded and pushed to the east. Sediment on the ocean side of the basalt pile, pushed from behind by the Farallon plate, underwent even more severe deformation. It was sliced and tilted and upended as it was shoved under the crescent-shaped rim of basalt. This sedimentary rock, once submerged off the coast, makes up the inner Olympics, the core of the range. In fact, the highest peaks in the Olympics (including Mount Olympus at 7,965 feet) are composed of former seafloor sediment. The sediments were crammed into the corner,

The highest peaks in the Olympic Mountains are composed of former ocean bottom.

thrust upward and then, as erosion lowered the surrounding terrain, became the range's highest peaks.

Ages of the sedimentary rocks in the core of the Olympic Mountains become younger to the west, as would be expected if the Farallon plate continued to subduct beneath the continent and add scraped-off sediment to the Olympics. Some of the Farallon's sediment cover was dragged beneath the growing Olympic Mountains. As this sedimentary rock jammed under the continent, it lifted the overlying mountains even higher. The most recently arrived sedimentary rocks can be seen along the Olympic Peninsula's beaches where visitors can walk over tilted and even overturned layers of former ocean bottom.

Apparently this process of jacking up the Olympics is continuing. Geologists believe that without continuing uplift, erosion caused by extraordinary amounts of snow and rain would have long since reduced the Olympic Mountains to rolling hills. (The spectacular peaks and valleys of the Olympics were caused, of course, by the erosive action of glaciers.)

The pushing and folding that overturned and bent the rigid basalts of the Crescent formation apparently created a valley that millions of years later became an ice-age channel for a huge glacier that moved out of Canada. The mile-thick ice sheet deepened and probably widened the valley; when the ice retreated, the valley was flooded with seawater, forming the Strait of Juan de Fuca between the Olympic Mountains and Vancouver Island.

❖❖❖

40 million years ago: volcanoes erupt from a sandy plain, marking the birth of the Cascade Range that separates the coastal and inland areas of the Northwest.

8

Separating East from West

Dozens of volcanoes erupted from a broad sandy plain with meandering streams along the Pacific Northwest's ancient coastline. Lined up for hundreds of miles, perhaps 50 volcanic centers with several volcanoes in each blasted away, sometimes simultaneously. It must have been one of the most spectacular volcanic displays the earth has experienced.

This was the beginning of the Cascade Range, the mountain barrier separating the moist west sides of Oregon and Washington from deserts to the east—perhaps the most dominant feature of today's Pacific Northwest.

The Cascades' violent birth began about 40 million years ago. Volcanoes rose and eventually died, eroding away and leaving only their roots for today's geologists to pick over. But as old ones snuffed out, fresh eruptions would take their place. Typically, these eruptions produced andesitic lava—the thick, sticky kind that tends to pile up into prominent mountains. Those eruptions, continuing over about 20 million years, built the foundation of today's volcanic Cascade Range extending from central Washington south through Oregon and into Northern California.

Setting the Stage

But before we get into the story of the Cascades, let's set the stage for the way things were in the Pacific Northwest 40 million years ago. There are several points to keep in mind:

- The coastline would have appeared only partly recognizable. Much of the British Columbia coastline and Vancouver Island were in place, having arrived millions of years earlier. But from the south end of Vancouver Island, the coastline ran southeast toward what is now central Idaho and then continued on south. The volcanoes that marked the birth of the Cascades rose along this coastline.

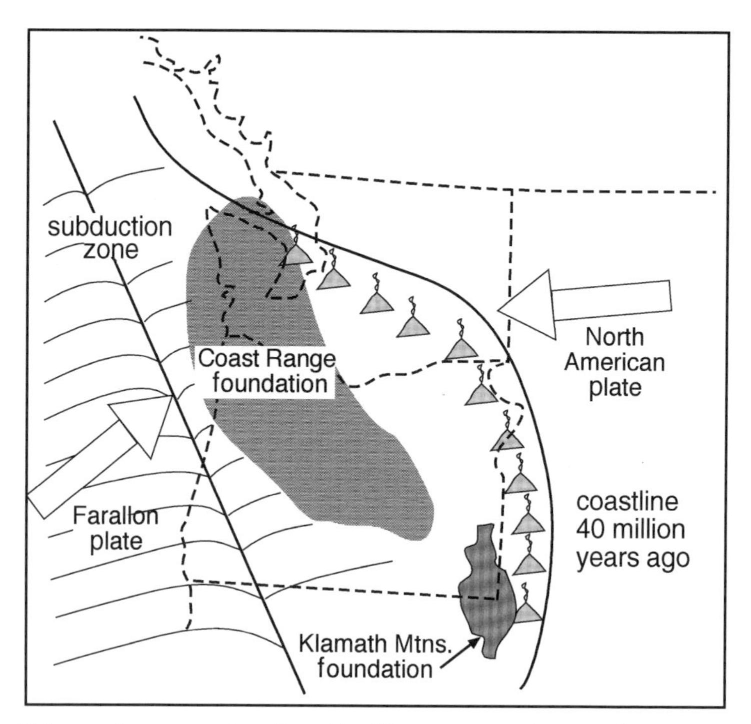

Volcanoes began erupting about 40 million years ago along North America's ancient coastline while the foundations of the Coast Range still lay offshore.

- The raw material for the North Cascades, British Columbia's Coast Range, and the San Juan and Gulf Islands was in place, although awaiting later uplift and much later glacial sculpting.
- Off the coastline was the great mass of basalt that forms the foundation of today's Coast Range in Oregon and Washington, including the Olympic Mountains. Forty million years ago, the rock was still submerged or barely above the surface.
- North America, creeping west, was still colliding with an ocean-bottom plate as it had for millions of years. But stresses exerted on the coastline had changed. The fast northbound Kula plate, which had broken off chunks from the continent and carried them north, was gone, subducted beneath Alaska. Its slicing action had been replaced by the Farallon plate's slower northeast push against the continent. The subduction zone—the trench where the Farallon plate bent down to slide beneath North America—was on the ocean side of the underwater basalt deposits. So pressure by the Farallon plate not only was squeezing the basalt closer to the coastline but was actually wrinkling and buckling the edge of the continent.

The Cascades Form in an Underwater Trough

One of these wrinkles—a deep trough—plays an important part in the story of the Cascades. A trough or valley had formed in the bottom of the shallow sea just off the coastline. As compression pushed the trough deeper, it began to fill with sediment from the swampy coastline and highlands farther east. The Cascade volcanoes formed along this trough 40 million years ago, although it is not clear why. It's possible the forces that pushed down the long underwater valley had left weak places in the earth's crust, providing a path for magma to rise from far below the surface.

As the volcanoes grew, at first erupting underwater, their weight pushed the bottom of the trough down even more. As the bottom continued to fall out of the trough, sediment from the continent and volcanic debris kept it filled. It was the beginning of what today is a pile of sedimentary rock at least three miles thick beneath the Cascades. Later uplift and erosion have exposed the edges of the pile in a few places, enabling geologists to calculate its thickness. It's interesting that the Cascade mountains we see are only the upper part of that massive foundation, the mere tip of the iceberg.

Subduction of the Farallon plate continued, sending great pools of magma pushing toward the surface, building a ridge that rose above the trough. Some of the magma erupted at the surface as volcanoes. Most cooled slowly a mile or so beneath the surface to form the batholiths, now exposed by erosion, that form a large part of the modern Cascades.

The Columbia Embayment

East of the rising Cascades was a low-lying basin, swampy and here and there invaded by arms of the ocean, a feature known today as the Columbia embayment. It covered today's Columbia Basin and the southwest part of Washington State. Much of the Puget Sound lowland and the Willamette Valley was still under water. The future site of the Cascade Range from Snoqualmie Pass in central Washington south through Oregon was partly submerged.

The Swinging Coast

Now we need to consider another part of the action that would perform major rearrangement of the landscape, something that had begun long before the Cascade volcanoes came to life: an actual swinging of the coastline outward from the continent. It was as if the coastline, along with the long volcano-filled valley, was swinging clockwise from a hinge somewhere in Washington's North Cascades. The rotation, as geologists call it, would eventually push the coastline out to its modern location with the Cascade and Coast Ranges in their familiar north-south orientation. But the movement was very, very slow, over millions of years.

It's hard to visualize: a huge segment of the west coast—volcanoes, swamps, the entire Columbia embayment—actually tearing loose and moving away from the continent. There is evidence that at times during the rotation the sea invaded the low area between the ancient continent and the detached coastline.

As the section of coastline swung from its "hinge," the foundation of today's Klamath Mountains began to move. The building blocks of the future Klamaths were the volcanic oceanic islands that had jammed against North America's west coast, which was still far to the east near today's boundary between Oregon and Idaho. The wreckage of former islands probably not much above sea level, they moved as much as 300 miles over millions of years as the coastline swung toward its modern location. By the time the Klamaths arrived at their present position, parts of the mountain range had rotated clockwise as much as 90 degrees. Today the Klamath Mountains occupy the southwest corner of Oregon and the northwest corner of California.

What made the coastline rotate?

There are a couple of likely explanations for what caused the rotation of the coastline.

Visualize the rigid ocean-bottom plate, a solid-rock slab several miles thick, striking the coastline at an angle. It doesn't glance off—it's too heavy, with too much momentum. Instead, it dives beneath the continent, exerting an oblique horizontal pressure. The edge of the continent has already begun to fracture and buckle. Under this slanting pressure, some of the fractures break open, allowing sections of coastline to swing out like a gate on a hinge.

Back-Arc Spreading

What geologists call back-arc spreading is also believed to have played a part. The arc refers to the line of volcanoes above a subducting plate. The back arc in this case was on the continental side of the volcanoes, opposite the subduction zone. When the sinking rate of the subducting plate is faster than the advancing motion of the overriding plate, the overriding plate may be pulled forward. The pull subjects the earth's crust behind the mountains to stretching and thinning, usually forming a basin. As the back arc stretches, the chain of volcanoes moves toward the subduction zone.

Eruptions continued over millions of years even as the growing Cascades were swinging away from the continent. During quiet times, which probably extended over centuries or thousands of years, erosion dumped sediment in low spots, forming soil which permitted forests to grow.

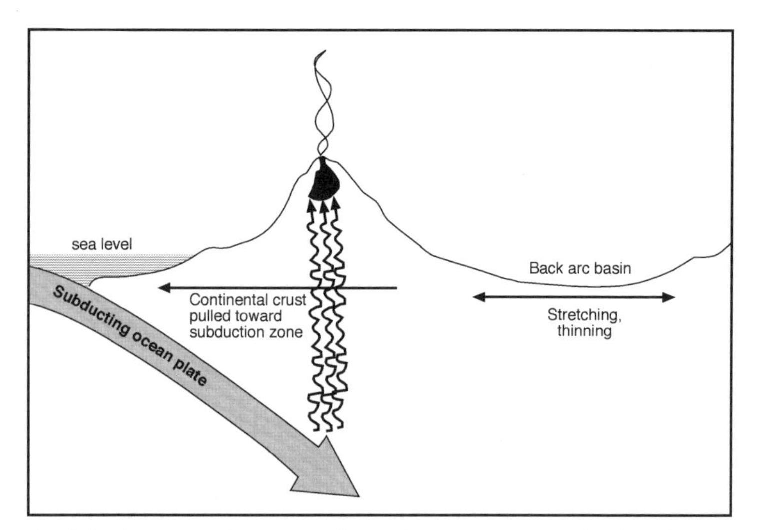

A subducting ocean plate may pull a continental plate toward it, forming and stretching a basin behind the volcanic arc.

But during active periods, it must have been spectacular: volcanoes extending over hundreds of miles blasting great clouds of ash into the atmosphere, avalanches of glowing-hot material racing down slopes at speeds approaching 100 miles an hour, and lava igniting vegetation as it filled valleys eroded since the last series of eruptions.

Debris from the volcanoes rode streams and the wind to find a natural resting place in the low, swampy basin east of the young Cascades. Volcanic ash and mud flows filled valleys and preserved fossils of plants and animals that indicate the climate was warm and moist, similar to that of the southeastern United States today. The older Clarno and John Day deposits in central and northeastern Oregon, famous for the well-preserved fossils they contain, were covered and preserved by volcanic sediment shed from the growing Cascades during this period.

The fossils of subtropical plants in those formations tell us something else: since the basin east of the mountains was wet and swampy, it means the Cascades were still not high enough to form a barrier to rainfall blowing in from the ocean, as today's Cascades do. Apparently, erosion and continued settling of the trough were keeping the mountains low.

On the ocean side of the volcanoes, volcanic debris buried enormous swamps. (Millions of years later, the compacted vegetation had formed coal, attracting miners to the western foothills of the Cascades.) Debris from the Cascades overflowed the swamps and covered the partially submerged portions of the great basalt flows that still lay offshore. Thus began the construction of the Coast Range of Oregon and Washington—a basalt basement covered by volcanic debris. Uplifting would come later.

The Cascades-building volcanism that began about 40 million years ago continued in three main stages until about 17 million years ago. Geologists call the deposits laid down during that 22-million-year span the Western Cascades, referring to the fact that these older formations are generally a little west of the younger, higher peaks of today's range. The Western Cascades, old and heavily eroded, make up the foundation beneath the much younger volcanoes we see today.

Floods of Fluid Rock

When the era of Cascade volcano eruptions ended about 17 million years ago, it was followed by an even more spectacular episode: massive outpourings of melted basalt so hot and fluid that it flooded the countryside almost like water. This lava, called basaltic lava, erupted from fissures or cracks that opened in the basin east of the Cascades, the area where back-arc spreading was stretching and thinning the earth's crust.

The flows of basaltic lava tell us something about the Cascades 16 million to 17 million years ago: even though the range was still too low to block rain clouds from the basin, it stopped the lava floods which pooled east of the mountains. There was one exception: an incongruous valley through the Cascades that several of the lava floods followed to the sea. This low point in the range, perhaps an old fault left in the earth's crust by the stresses and strains of rotation, was a wide valley located about where Oregon and Washington meet today. We recognize this as the gap where the Columbia River crosses the Cascades. And, indeed, the ancestral Columbia also utilized that route to the sea; this unusual valley persisted through millions of years, even as the Cascades were rising and rotating.

The huge plumes of lava that flowed hundreds of miles from their source tell us more about the rotation that swung the Cascades. The last great basalt flood to cross the Cascades, the Pomona flow, occurred about 12 million years ago. Paleomagnetic research reveals that the end of the Pomona lava flow near the ocean has rotated clockwise 16 degrees in 12 million years, as compared to the rest of the same flow east of the Cascades.

A Time of Volcanism and Uplift

The young Cascade Range had been relatively quiet for about 8 million years during the time of the great basaltic lava floods east of the mountains. But as the lava floods decreased, the Cascades entered a stage of activity that continues today—volcanism and uplift. The growth of the Cascades from a range that was too low to trip up rain clouds has occurred largely in the past 9 million years. The renewed eruptions produced andesitic lava—the rubbly, explosive kind that builds mountains—unlike the fluid basaltic lava that had flooded the basin.

It's tempting to look for a connection. As far as geologists can tell, the Cascade volcanoes pretty much shut down, or at least changed their pattern of eruptions, during the millions of years that basalt was covering the basin to the east. The mountain building resumed as the basalt floods were winding down. It almost certainly had to do with interactions among the tectonic plates: North America creeping west, the Pacific plate pulling to the northwest, and the smaller Juan de Fuca plate (a remnant of the old Farallon plate) pushing northeast as it dived beneath the continent. But it's difficult to pin down cause and effect.

There are some fascinating hints—or maybe coincidences. About 10 million years ago, when the great basaltic lava floods east of the Cascades had just about come to a halt, the Juan de Fuca plate began slowly rotating clockwise, as indicated by clues in the ocean bottom. Its direction of movement changed from north-northeast to northeast, making its collision with North America more head-on, less of a glancing impact.

This could have had two effects: First, it would have lessened the shearing movement that may have been helping to pull the Cascade and Coast Ranges away from the continent. Second, the change in the Juan de Fuca plate's motion would have increased the rate at which the subducting plate was diving beneath the continent. This could have sent pools of magma rising into the old Western Cascades, pushing them higher and re-igniting a line of volcanoes along the ridge.

How much have the Cascades risen?

The big flows of basaltic lava east of the range provide an unusual way to measure the uplift of the Cascade Range throughout its most recent geological phase. Geologists use the flows as a huge leveling device.

We know some of the flows spouted from fissures at an elevation of about 2,500 feet along what today is the Washington-Idaho border. We know that the hot, thin lava flows must have been flat when they cooled, like big lakes. And we know that the flows lapped up against the still-low Cascades far to the west, covering some of the foothills. Today there are remnants of those basaltic lava floods high in the Cascades of Washington, indicating uplifts of at least a mile. Basalt high in the Blue and Wallowa Mountains of Oregon indicate similar uplift. Most of it has occurred in the past 9 million years.

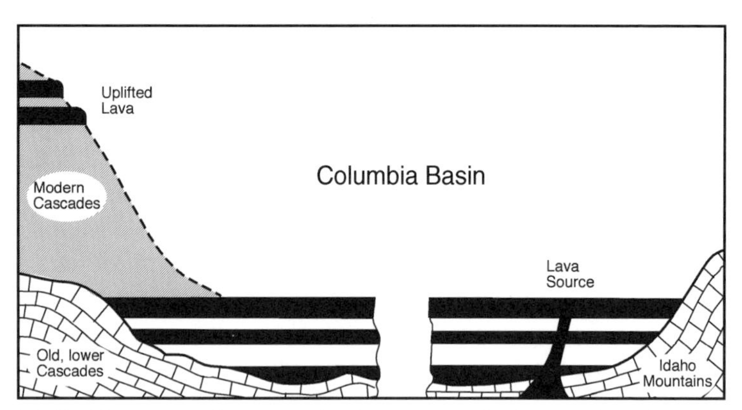

Uplifted fragments of ancient volcanic rock layers indicate the Cascades have risen as much as a mile since the lava flowed.

The best place to see effects of this uplifting of the Cascades is in the Columbia Gorge; here the river maintained its cross-Cascade route even as the mountains rose. Basalt layers formed by repeated flows were pushed up into an arch, with the highest point near Bonneville Dam. Uplift in the gorge, the lowest part of the Cascade Range, is 2,800 feet—about half that near Mount Rainier.

The High Cascades Period in Oregon

The floods of basaltic lava in the Columbia Basin had almost ceased when the Cascade volcanoes roared back to life, beginning what geologists call the High Cascades period. This period of volcanism, less intense than the era of eruptions that built the Western Cascades, continues today. High Cascades volcanoes rose atop the old Western Cascades foundation, died, eroded away, and were replaced by more volcanoes. The snow-covered volcanoes we see today are as temporary as those that preceded them. They are known to geologists as stratovolcanoes, built of easily eroded layers of lava, ash, and clay.

There is evidence that the Cascades are continuing to rise. One clue is provided by a volcano that erupted about 2 million years ago southeast of present-day Mount Rainier. Most of that ancient volcano is gone, eroded down to its rocky roots in the area we know as the Goat Rocks Wilderness. The Goat Rocks volcano was probably a little smaller than Rainier, but in its youth it sent a stream of thick andesitic lava 40 miles to where Yakima's outskirts are today. It's the longest andesitic flow known in the world.

Much of the way, the plume of lava followed the canyon of the Tieton River, providing scientists with a unique method to measure tilt. By a complicated process, one geologist has come up with a preliminary estimate that the slope of the Tieton's canyon today is three times steeper than it was when the Goat Rocks lava pushed through it. That indicates the upstream end of the river canyon, along with the Goat Rocks deposit, has been rising with the Cascades since the lava flowed 2 million years ago.

Not Your Usual Volcanoes

When we think of volcanoes in the Cascades, we tend to think of the snow-capped peaks along the skyline in Washington, Oregon, and Northern California. But, surprisingly, those volcanoes account for only about 25 percent of the lava that flowed during the High Cascades period. The flows that produced the most volume were of a very hot, fluid basalt that didn't pile up into peaks. Instead basaltic flows built gently sloping shield volcanoes, covered lowlands with pools of lava, and spit up ash to build cinder cones around vents. Most of the basalt activity was south of the Columbia River where the earth's crust was still being stretched and thinned as the mountains rotated clockwise.

Newberry Crater, a shield volcano 25 miles south of Bend, is Oregon's biggest volcano—with a greater volume than either Mount Hood or Mount Jefferson. It began erupting about 500,000 years ago and built a huge, sloping pile of ash before it exploded violently and collapsed into itself, forming a crater or caldera. Newberry Crater produced a huge ash eruption 1,600 years ago and has been active as recently as 1,300 years ago. There is a significant flow of obsidian in the crater. The obsidian, or volcanic glass, which can be flaked into edges sharper than steel blades, was used by local Indians and traded to tribes all over the Pacific Northwest.

Newberry Crater is representative of the back-arc volcanism that occurred during the High Cascades period. The volcano is 40 miles long and 20 miles wide, with a 4-mile-wide crater at its summit. Its slopes are sprinkled with cinder cones, some of which are in straight lines, indicating they formed along fractures or faults caused by stretching.

Lava Butte, about 10 miles south of Bend, is a 500-foot-high cone that rose along a fracture extending out from Newberry Crater. Erupting about 6,000 years ago, it sent lava flows for miles, damming the Deschutes River and covering a pine forest, forming today's Lava Cast Forest.

Congress recognized the importance of volcanism in shaping central Oregon when it established the Newberry Crater National Monument. It includes Hole-in-the-Ground, an explosion crater blasted out when molten rock encountered groundwater, creating an explosion of steam. Nearby is Fort Rock, a bowl-shaped formation composed of expelled ash that settled around the vent and cemented into a rock known as tuff. Thousands of years later, an ice-age lake eroded an opening into the bowl. Some of the earliest human inhabitants of the Northwest found shelter in shallow caves eroded into Fort Rock.

The Explosion that Formed Crater Lake

One of the more remarkable events of the High Cascades period was the catastrophic explosion of Mount Mazama in southern Oregon. Born about 400,000 years ago during an ice age, Mazama grew to an estimated elevation of 10,000 to 12,000 feet before it exploded violently about 6,900 years ago. It sent up glowing hot ash into a towering cloud that was distributed by winds over much of the Pacific Northwest.

Almost immediately after the blast, the Mazama volcano collapsed, removing an estimated 2,000 feet from the summit and forming a caldera more than five miles across. As the mountain caldera cooled, rain and melting snow formed what is now known as Crater Lake. The lake's Wizard Island is a cinder cone built by a later eruption in the caldera.

Oregon's stratovolcanoes—Mount Hood, Mount Jefferson, Three Sisters, Diamond Peak, Mount McLoughlin—grew atop the older Western Cascades. In Northern California, Mount Shasta and Lassen Peak rose during the same period.

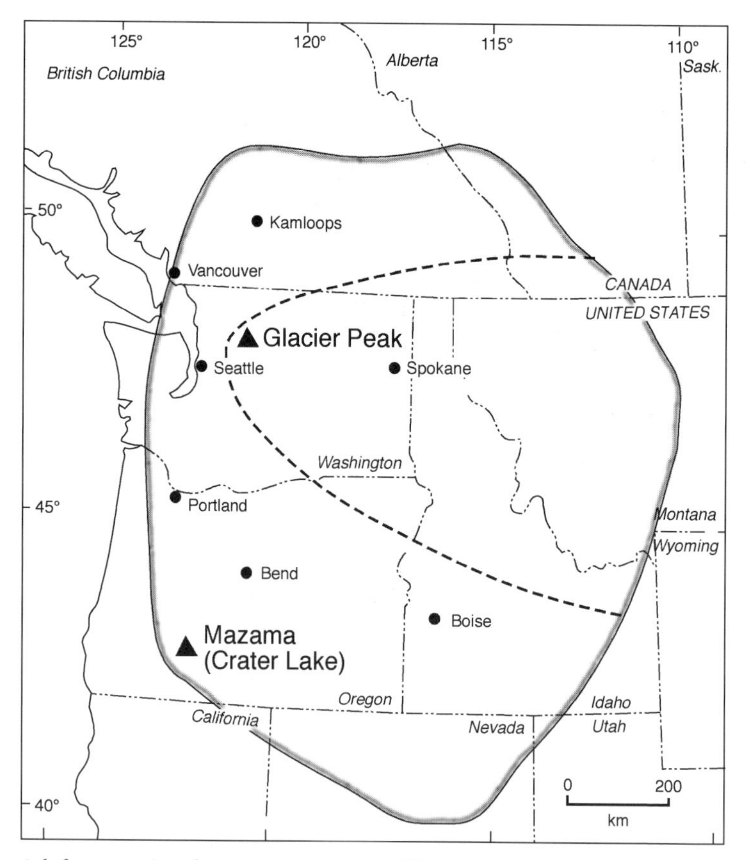

Ash from a series of enormous eruptions of Mount Mazama about 6,900 years ago settled on most of the Pacific Northwest, indicated by the shaded line on the map. Glacier Peak erupted about 12,000 years ago and sent an ash plume east and southeast, shown here as the area enclosed by dotted lines.

Volcanic Ash as Time Marker

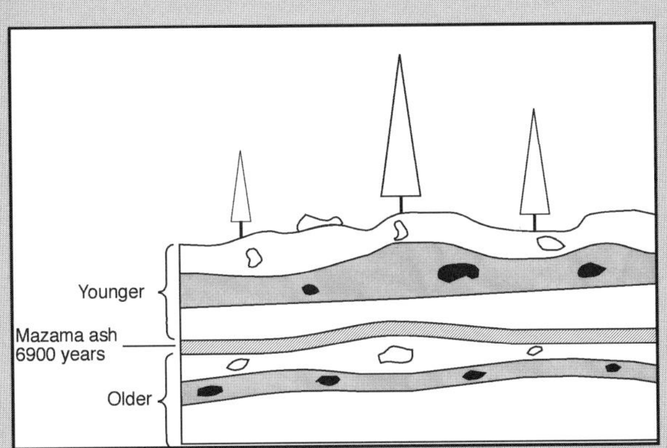

The wind must have been out of the southwest when Mount Mazama erupted in a series of explosions, sending up a towering cloud of ash that settled over much of the Pacific Northwest. Seventy feet thick on Mazama's slopes, the ash lay several feet deep for miles downwind and gradually tapered off with distance. The ash, visible as a gray streak buried beneath younger, darker sediment, is found in most of Washington, Oregon, and Idaho, and in parts of Montana, Nevada, California, British Columbia, and Alberta.

Radioactive dating of the ash reveals that Mazama blew up approximately 6,900 years ago. The widely distributed layer of ash has become a valuable time marker for scientists in many fields, including paleontology, archaeology, and geology. Simply put: anything found below the ash is older than 6,900 years; above is younger. The Mazama ash has even been useful to oceanographers studying the ocean floor off the Oregon-Washington coast where ash was deposited by the Columbia River.

Humans had been in the Pacific Northwest for thousands of years when Mazama blew. Signs of human habitation have been found below a layer of Mazama ash in several archeological sites in the Pacific Northwest, including caves in Fort Rock—just 50 miles northeast of the mountain. The ash, which would have sifted out of the atmosphere for weeks or months, must have added considerable difficulty and danger to these early people's lives, affecting everything from the weather to the availability of food.

Another useful time marker, even older than Mount Mazama, dates from 12,000 years ago, about the time humans were arriving in the Pacific Northwest. This older ash deposit came from the eruption of Glacier Peak in Washington's North Cascades. The enormous load of ash was carried toward the east and southeast, dusting parts of eastern Washington and Oregon, northern Idaho, and western Montana.

Scientists can tell apart Mazama ash and Glacier Peak ash by their differing chemical and mineral composition, making these two large ash deposits extremely useful for dating geologic events throughout much of the Pacific Northwest.

The High Cascades Period in Washington

There was less volcanic activity in Washington than in Oregon during the High Cascades period, perhaps because the mountains north of the Columbia River were closer to the "hinge," with less rotation and back-arc spreading. The Indian Heaven volcanic field in southern Washington spewed huge volumes of lava on and off for about a million years. The last eruption, 8,150 years ago, produced the Big Lava Bed. This volcanic field erupted from seven low, gently sloping cinder cones. (Interestingly, two of the cinder cones erupted beneath ice age glaciers.) By the time Indian Heaven was exhausted, Mount Adams to the east and Mount St. Helens to the west had built stratovolcanoes that now tower over the low-lying volcanic field.

The Simcoe volcanic field erupted on the east slope of the Cascades in central Washington around 4 million years ago. It lies atop two older deposits, the Ellensburg formation, formed by sediment washed down from the growing Cascades, and the Columbia River basalt formation, which had erupted far to the east.

In Washington, the High Cascades period produced Glacier Peak and Mounts Baker, Rainier, Adams, and St. Helens as a result of subduction of the Juan de Fuca plate. Mount Baker and Glacier Peak were built by magma rising through thousands of feet of former ocean bottom that had arrived as part of Superterrane II, 100 million years ago; the North Cascades, Rainier, Adams, and St. Helens were built atop younger rock of the Western Cascades. Characteristics of the lava in each mountain is distinctive, reflecting the different environments through which the magma passed.

Mount Rainier, Washington's oldest volcano of the High Cascades period, was born about 500,000 years ago. Mount St. Helens, which awoke in 1980 after more than a century of dormancy, is the youngest, at about 40,000 years.

As subduction of the Juan de Fuca plate compresses the edge of the continent and sends magma rising beneath the Cascade Range, uplift continues today. If it had stopped, rapid erosion from glaciers, melting snow, and rain would have begun rounding off and lowering the range.

One estimate is that the Cascades rose about 2½ inches per 1,000 years until about 4 million years ago, and less since then as the rate of subduction slowed. With the Cascades rising 2 inches every 1,000 years, or 2/1,000ths of an inch a year, even the new satellite-based Global Positioning System would need a century or so to detect the change.

❖❖❖

17 million years ago: volcanism erupts east of the young Cascades, beginning one of the world's greatest outpourings of basalt.

9

A Deluge of Lava

At first, the changes must not have amounted to much: a few small earthquakes; cracks that appeared in the ground like parallel stretch marks. It was in what might be called a backwater of the evolving Pacific Northwest, a place where very little had happened for millions of years. We know it today as the Columbia Basin, the part of Washington and Oregon bordered by the mountains of Idaho to the east and the Cascade Range to the west, the Okanogan Highlands to the north, and the Blue Mountains to the south. But 17 million years ago, the Cascades were still low enough that clouds from the ocean delivered plenty of rain to the swampy basin. Rivers from the surrounding highlands drained into the basin which, in fact, was not much changed from its days as a shallow estuary of the sea.

The climate was warm, which made the basin a good place for plants and animals. Primitive horses, about the size of donkeys, grazed on grassy hills among oak, maple, sweet gum, and cherry trees. Sometimes they retreated as elephant-like mastodons moved near to tear leaves from the trees. An occasional camel padded along the hills. In the valleys, rhinoceroses wallowed in shallow water. Giant beavers felled willow and cottonwood trees

A Window to the Past

Department of Energy photo

Rattlesnake Ridge, part of the Yakima folds, was selected for exploratory oil and natural gas drilling in 1957. The ridge, where basalt layers had buckled and thrust upward, theoretically held promise as an underground trap where hydrocarbons formed from ancient swamps might be found.

The drill penetrated nearly 100 individual basalt flows, but when the drilling reached a depth of 10,655 feet and still had not broken through the basalt, the geologists gave up and capped the well. Although a bust for the oil company, the dry hole was a boon to geologists—proof of the stupendous thickness of the basalt where flows had pooled over a sagging basin floor.

The well remained capped for almost 10 years until geologist Randall Brown suggested utilizing it for scientific exploration. The Atomic Energy Commission, at that time considering a plan to dispose of radioactive waste in caverns dug deep into the basalt, put up money to provide "unique access to one of the world's thickest basalt sections." The two-mile-deep drill hole re-opened in 1967.

The drill hole revealed much about how different the Columbia Basin had been millions of years ago. Analysis of pollen preserved in soil between the layers of basalt showed that warm-climate plants like ferns and mosses, cypress, sweet gum, beech, and hickory trees once flourished here. The evidence pointed toward a warm, humid climate with at least 40 inches of rain a year, where droughts were rare and even winter temperatures remained above freezing. Pollen from higher-elevation trees was also found, including hemlock, larch, and Douglas fir; scientists assume this pollen was blown from highlands to the west where the Cascades were rising. They found no evidence of a rain shadow, indicating the Cascades were still too low to block storm systems from the west.

Today the area gets as little as 7 inches of precipitation per year. Native vegetation includes bunch grass and sagebrush. The weather can be blazing hot in summer and freezing cold in winter. The abandoned drill hole confirmed earlier evidence that what is now a dry sagebrush steppe was once a subtropical swamp dotted with lakes.

to dam meandering streams. They shared their ponds with three-foot-long turtles.

The verdant scene was about to change. As the small earthquakes increased in number, the cracks in the ground grew more prominent. If modern instruments had been in place they would have shown that the ground was being pushed up and tilted away from the cracks. The cause? The ground in the eastern part of the basin was actually being pulled apart, opening cracks far below the surface. Streams of magma were pushing their way up through the narrow seams, lifting and cracking the surface.

Over several months, the seismic action would stop occasionally, only to begin again. Eventually, the earthquakes grew so numerous they overlapped; today scientists know that such tremors are caused by magma on the move.

One day, or perhaps it was night, low-hanging clouds glowed orange, reflecting incandescent light from molten rock rising in the widening cracks. Within hours, red-hot lava burst in fiery fountains from the cracks. The fountains spread along the cracks to form sheets of fire leaping hundreds of feet high, extending for miles. One of the most stupendous events in the evolution of the Pacific Northwest had begun.

Columbia River Basalts

Over millions of years, one gigantic lava flow after another burst from long fissures to send lava over vast areas. The basaltic lava was so hot and thin it flowed freely, enabling it to travel hundreds of miles. These flood basalts in the Columbia Basin, known collectively as the Columbia River basalts, comprise the second largest such floods known to have occurred anywhere in the world, exceeded only by flood basalts in India about 60 million years ago. And some of the individual flows of Columbia River basalt were the largest known flows on land anywhere, or at any time. One of the biggest flows, now spread over much of the basin, contains about seven times the volume of Mount Rainier. And it probably erupted over just a few days, or perhaps a few weeks.

More than any other geological event, the lava flows shaped the Columbia Basin as we know it. Repeated flows, stacked one on top of another, total more than two miles thick in the central part of the basin. Rivers have cut into the upper layers of the stack, revealing the layers in canyon walls. Much later in the story of the Northwest, great floods from melting ice sheets left cliffs, mesas, and jagged outcrops of the basalt flows that once had been as flat as pancakes.

The great flows originated in the area where the states of Washington, Oregon, and Idaho now meet. The erupting fissures were aligned in a northwest-southeast direction, at right angles to the forces that were stretching the surface.

The first flows would have leveled the old marshy basin, filling valleys and river canyons and covering the lower hills to leave a flat surface shimmering with heat. Subsequent flows would have traveled faster and farther, taking advantage of the leveled surface. A few hills, such as Steptoe Butte in eastern Washington, were high enough that their peaks remained above the lava flood. Steptoe Butte, in fact, lent its name to science: geologists use the word steptoe to denote a peak rising above a younger lava flow.

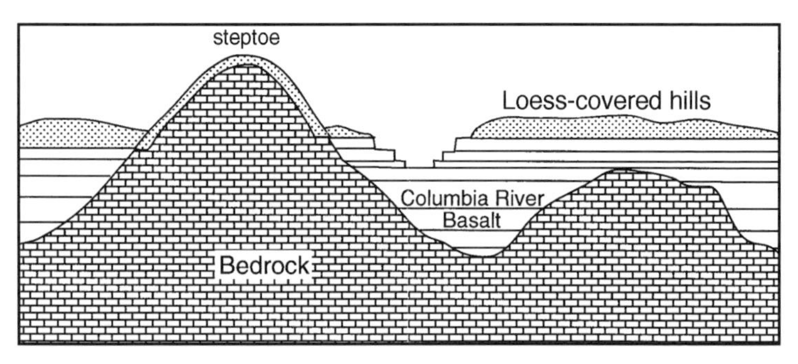

An ancient peak protruding above younger flows of lava is known as a steptoe, a name derived from Steptoe Butte along the Washington-Idaho border.

How big were the lava flows?

Trying to visualize one of those great flows is mind-boggling. Imagine a lava flow perhaps 100 feet high and more than 50 miles wide. On level ground, it advanced at 3 to 4 miles per hour, equivalent to a brisk human walk. On a downslope, the speed picked up to around 10 miles per hour. The molten rock, at 2,000 degrees Fahrenheit, instantly ignited trees, grass, underbrush—anything it approached. As the smoking gray-black mass plowed into lakes and swamps, the water exploded into steam. As the lava pushed on, only steam jetting through a still-hot lava sheet marked where a lake had been.

Relics of the Flows

The Blue Lake Rhino—As a basalt flow plowed into a marshy pond, it covered the body of a small rhinoceros floating there, probably killed by the heat and volcanic gasses. The lava, cooled by the water, encased the rhino's body, making a crude mold. The mold was found in 1935 in an eroded cliff above Blue Lake in the Grand Coulee of central Washington. Only a few teeth and bone fragments of the rhino survived, but the shape of its body was perfectly preserved by the lava mold. A replica of the lava mold that encased the Blue Lake Rhino's body can be seen at the Burke Museum at the University of Washington in Seattle.

Petrified Wood—After a pause between eruptions long enough for soil to form and forests to return, a new lava flow overran swamps and covered waterlogged trees. Over millions of years, minerals replaced the wood to form stone that preserved the cell structure of the wood, even the growth rings. Later as earth movement pushed up ridges, the ancient logs were exposed and are now found in central Washington along the Columbia River. The Ginkgo Petrified Forest State Park near Vantage was named for the ginkgo tree, one of the many species identified there; today the ginkgo tree occurs naturally only in East Asia.

Boulders blown from the sides of the erupting fissures flew shrieking a mile or more before crashing to the ground. There must have been a continuous roar as the fire fountains explosively heated the air. The heat plume triggered lightning and booming thunder. Ground-level air rushing into the vacuum created by the rising plume spawned howling winds that knocked down trees in the direction of the glowing fountains. Felled trees touched by lava exploded into flame.

The outpouring would have been visible from the moon if anyone had been there to see it. By night, it was an angry scar on the darkened earth, a glowing sheet spreading from a line of fire. By day, smoke and ash clouds sent plumes around the world. Paradoxically, heat and gasses from the eruption probably cooled the earth by increasing cloud cover.

In lesser eruptions, the molten rock would have piled up into gently sloping shield volcanoes, as eruptions of somewhat similar basalt do in Hawaii today. But the rate of production from the larger erupting fissures—greater than humans have ever witnessed—simply pushed the lava outward in great sheets. During one of the bigger flows, produced by a 60-mile-long system of fissures, just one day's output exceeded the total volume of Washington's Mount St. Helens, even before it lost its top in 1980.

How far did the lava flow?

Because mountain building in Idaho had tilted the Columbia Basin slightly, most of the basalt moved west and northwest. The early flows followed the canyon of the ancestral Snake River, filling it and forcing the river to find a new route. Later flows coasted over the flattened basin to lap up against the Okanogan Highlands to the north, pushing the ancestral Columbia River out of the basin and up against the highlands.

The flows also reached the edge of the Cascades which, although still not much of a mountain range, diverted the lava along foothills to the south. There the lava floods found a broad valley that cut entirely through the young mountains—the same valley that the ancestral Columbia River had followed for millions of years as it drained the swampy interior basin.

The bulldozing front of the arriving lava flow shattered the peaceful river valley. The spreading basalt sheet probably had disrupted the flow of the Columbia far upstream. But if there still was water in the cross-Cascades valley, it would have exploded into steam as the molten rock forced the river out of its channel.

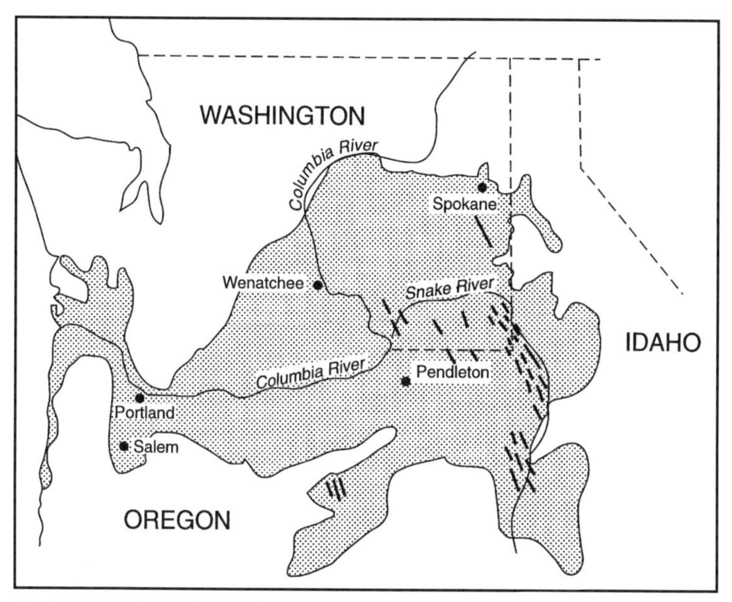

Shaded area shows the combined extent of flows of the Columbia River basalts over millions of years. Short lines indicate fissures from which lava erupted.

The basalt poured down the river valley, coursing across what is now northern Oregon, flowing south of the present channel of the Columbia and over the site where Mount Hood would rise millions of years later. Lava emerging from the cross-Cascades valley also flooded southwest Washington, leaving deposits as far north as Grays Harbor. (There are a few puzzling deposits of Columbia River basalt located west of Mount Adams. A separate flow may have found another path through the Cascades; if so, geologists have not figured out how this occurred.)

Lava flowed south in western Oregon, reaching beyond Salem in the Willamette Valley. Along the way, the lava flood found passages through the young Coast Range (at that time still a discontinuous series of hills) to reach the ocean. On the coastline, it burrowed under lighter sediments, floating them like rafts. It's still uncertain how far offshore the tunneling lava reached, but by the time it stopped, it had traveled at least 450 miles from the fountaining vents.

New Routes to the Sea

When each outpouring of lava ceased, the Columbia River did what it had to do: find a new way to the ocean. Repeatedly, the river moved a little north of its old channel, which was plugged solid with hardened basalt. And when the next great outpouring of glowing lava arrived, it followed the "new" river channel to the ocean. The result was a series of basalt-filled channels crossing Oregon.

Later volcanism and uplift of the Cascades complicate the tracing of the old river channels along their entire length, but the places where they reached the ocean are obvious. As each basalt flow approached the coastline, it sank into the soft bottom of an estuary, filling successive mouths of the ancestral Columbia River. The lava cooled where it had burrowed deep beneath coastal sediment. As the coastline rose over millions of years, wave action eroded the sediment and exposed the tongues of basalt until they towered over the coast. Today they form striking capes and headlands both north and south of the mouth of the Columbia River. The major capes—from Seal Rock and Yaquina Head in Oregon

Cape Disappointment, topped by a lighthouse, is the relic end of a massive lava flow that originated hundreds of miles to the east. The cape rises above Fort Canby State Park in Washington, just north of the Columbia River.

to Cape Disappointment in Washington—are the relics of basalt flows that came out of the fissures hundreds of miles away.

At first, geologists thought the Oregon coastal basalts must have erupted in place. But the basalts, even though extending deep into ocean sediments, are "rootless." In other words, there is no source vent beneath them; it is obvious they must have arrived from someplace else. As more sophisticated techniques became available, geologists were able to link each coastal headland to a specific Columbia River basalt flow.

The time that it took the lava to flow from the fissures to the sea, located hundreds of miles away, appears to have been no more than a week or so. Even the farthest-traveled lava solidified into obsidian when it plunged into the sea and cooled suddenly. This indicates that the lava still was very hot and had had little time to cool as it traveled cross-country.

One Hundred Floods

There were more than 100 basalt floods over about 10 million years, although only 25 or so of the biggest ones reached the sea. There were smaller basalt flows until about 6 million years ago but they did not cross the Cascades. Thus, by 10 million years ago or so the Columbia River had settled into its present channel. Later uplift of the Cascades arched up basalt deposits, forcing the river to cut down through the rock to maintain its channel. This created the spectacular Columbia River Gorge between Oregon and Washington.

The first basalt floods, about 17 million years ago, erupted from vents in northeast Oregon. One million years later the feeder vents had moved to southeast Washington, and the volume of the flows increased dramatically. Geologists estimate that more than 85 percent of the total volume erupted during the first 3 million years.

Sometimes one flow covered another so quickly—after only a century or two—that there was not enough time for soil layers to form. But there also were long pauses between flows. Even during the time of maximum activity, flows averaged one for every 10,000 years. Sometimes there would be a 100,000-year pause. Then streams would bring sediment from the mountains, soil would form, vegetation took root, and erosion carved valleys and hills. But each time, the soil would be covered by a new flow that would burn the trees, explode lakes and ponds into steam, and lay down a flat layer of lava that would take 30 years or so to cool. Today, thin layers of ancient soil now compressed into rock can be seen between basalt layers in river-carved canyon walls, such as the Snake River canyon.

In the first century or so after each flow, the Columbia Basin must have been a wide plain, featureless except for an occasional ridge or peak protruding above the flow. The rock surface,

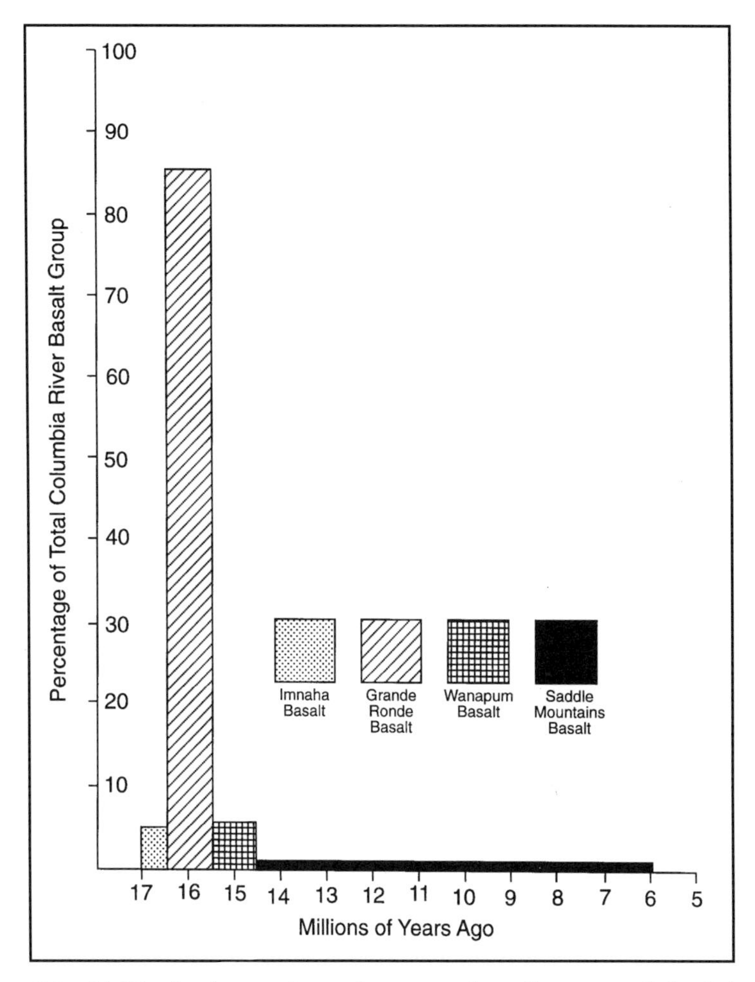

Columbia River basalt group. Stupendous outpourings of lava repeatedly flooded the basin east of the young Cascade Range beginning about 17 million years ago. The flows, known collectively as the Columbia River basalt group, occurred in four stages separated by long quiet periods when soil formed and vegetation flourished. The Grande Ronde basalt, including many individual flows, was by far the largest, totaling more than 600 times the volume of Mount Rainier.

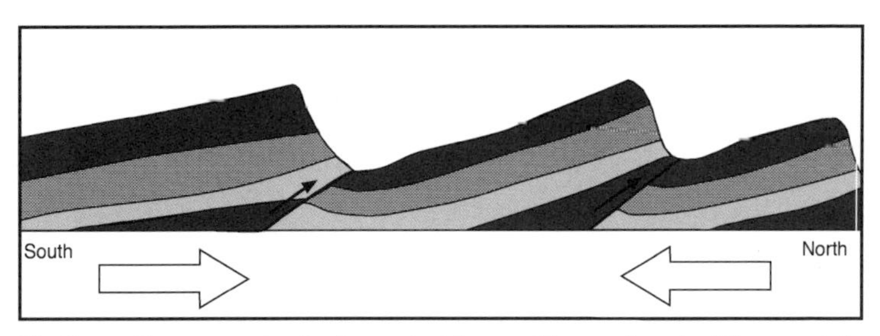

North-south compression broke ancient layers of lava and pushed them up into ridges, creating today's Yakima folds in central Washington. The ridges typically drop sharply on the north side.

still without soil cover, would have been nearly bare of plants. Until stream channels developed after thousands of years, rain water must have run off randomly toward the Columbia River where it had been pushed into the crease between the lava plain and the surrounding highlands.

The Yakima Folds

Over time, the hardened flows didn't maintain their flat-as-a-pancake shape. A north-south pinching accompanied the east-west stretching that opened the basalt-spewing fissures. Eventually the enormous pressure buckled the thick, rigid layers of basalt, pushing up ridges that run east to west. In some of these ridges, geologists have found layers tilted on end and, in some cases, overturned.

The north-south compression must have been occurring when the first flows flooded the basin. Remote sensing shows ridges in some of the earliest flows, which now are deeply buried. Later flows covered those fractures but broke at the same weak spots to again push ridges above the plain. The result is a system of east-west ridges in central Washington, known collectively as the Yakima folds, rising at right angles to the Cascades. They include the Saddle Mountains, Frenchman Hills, Horse Heaven Hills, Rattlesnake

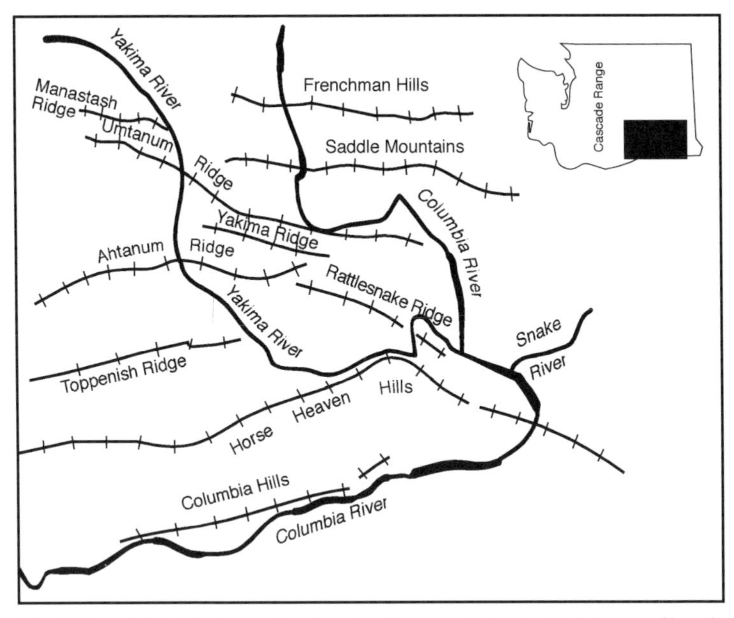

The Yakima folds. Not long after floods of lava cooled into thick layers of basalt east of the young Cascade Range, north-south compression pushed up ridges as the layers of lava broke under the pressure. The ridges are known today as the Yakima folds, which run more or less east-west, and perpendicular to the Cascades. Small earthquakes indicate that compression continues to push the ridges higher. The Yakima folds usually slope up gradually on the south side and drop sharply on the north, the side where rock layers broke and were pushed atop others.

Ridge, Yakima Ridge, Toppenish Ridge, and several other ridges. The Columbia Hills, along the north bank of the Columbia River, are part of the same fold system.

The direction of movement for small earthquakes in the Columbia Basin indicates that the north-south compression continues today. Some of the ridges appear to still be rising. Part of the Saddle Mountains, for instance, is believed to be rising at a rate averaging about one-sixth of an inch in 100 years. That's roughly equivalent to the width of a human hair every year.

Highway cuts through these ridges sometimes reveal how the layers of lava were arched up by the forces of compression. One such series of road cuts is on Interstate 82 where it transects the Yakima folds between Ellensburg and Yakima in central Washington.

Although details vary, most geologists agree the forces that pulled open the erupting fissures and pushed up the ridges were caused by interactions between the northwest-pulling Pacific plate, the west-creeping North American plate, and the subducting Juan de Fuca plate. As noted earlier, a subducting ocean plate can cause stretching, or back-arc spreading, in the part of the continent above the diving plate.

The accumulating weight of 100 lava flows pushed the old Columbia Basin even deeper. The flows ponded deepest at the low spot near the U.S. Department of Energy's Hanford Site at Richland, Washington. This mass of basalt is more than two miles thick. The top level of the youngest basalt flow in the center of the Columbia Basin—once located at ground level—has sagged 200 feet below sea level. It is covered by sediment delivered by wind and water.

The warm, swampy basin that preceded the basalt flows would seem to have fulfilled conditions favorable for creating oil and particularly natural gas. Basalt and sediment, especially where pushed up into ridges, provides an impermeable cap to trap and accumulate oil and gas. But so far, oil companies drilling into the basalt ridges have found only small quantities of natural gas, not enough to warrant commercial production.

❖❖❖

18,000 years ago: vast sheets of ice advance from British Columbia as far south as Olympia, Washington.

10
The Ice Age Arrives

One summer, little snow melted in the high mountains. The next summer was the same, and the next, until snow began to pile up on the peaks. After many years, it became compressed into ice and started to creep toward the valleys. As the climate continued cooling over thousands of years, glacial fingers probed downhill, bulldozing soil, rocks, and trees, even damming up rivers. The ice plowed and plucked and ground its way wherever the terrain permitted.

Glaciers spread over much of the cooling planet's Northern Hemisphere. A deep mantle of ice covered the mountains of British Columbia, from the Rockies to the Coast Range and the mountains of Vancouver Island. Glaciers found their way into valleys, filled them, and kept moving ahead. As they left the valleys, the ice spread out into thick sheets. The leading edges of the ice sheets could be messy, chaotic places—walls of ice several hundred feet high pushing mounds of debris into temporary lakes that had formed as the ice blocked rivers and rode over forests. Floods of muddy meltwater occasionally burst from beneath the glaciers.

By 18,000 years ago, ice covered the valleys and plateaus of interior British Columbia. It choked the Georgia depression, a huge valley between the Canadian mainland and Vancouver Island. Only the highest peaks poked above the ice. And by 16,000 years ago, geologists say, the broad ice sheet was pushing south across the forty-ninth parallel into what is now the United States.

As the massive white sheet moved south, it scooped out troughs that one day would fill with water to create beautiful lakes. Even before the ice sheet arrived in the lowlands, mountain glaciers had begun sharpening the peaks and scooping deeper valleys in the mountain ranges of British Columbia and northwest America, creating today's world-class mountain scenery.

A Gigantic Cleaver

The encroaching Ice Age brings us to the final stage in our story of the shaping of the Pacific Northwest. The advance and retreat of the ice left our corner of the country pretty much as we see it today. The British Columbia Coast Range and the Olympics and Cascades of Washington were in place when the ice arrived; they were about as high as they are today, but lacked sharp peaks and plunging valleys. The Coast and Cascade ranges functioned as a gigantic cleaver that split the massive Canadian ice sheet, sending a tongue west of the mountains to scoop

What Causes Ice Ages?

The ongoing process of plate tectonics, which has been pushing continents around for millions of years, plays a part in determining where glaciers and ice sheets form. The slow shifting of continents may place land areas near polar regions where there are oceanic sources of moisture, and, in continental collisions, push up mountain ranges high enough for snow to collect.

There is evidence that India and South Africa were glaciated 300 million years ago when they were adjacent to Antarctica as part of an ancient supercontinent. But it takes more to make an ice age than merely placing continents in polar regions near a source of moisture. Global cooling is also required; scientists believe such cooling can be caused by seemingly minor changes in the earth's orbit around the sun, combined with changes in the planet's tilt as it spins. These changes apparently can cause variations of as much as 10 percent in the amount of the sun's heat that reaches different parts of earth. Astronomers have reconstructed these orbital variations and they seem to match the history of ice ages during the past million or so years.

out and deepen the Georgia depression and the Puget lowland, which then was a broad valley drained by north-flowing streams. East of the Cascades, the ice sheet had slower going where it encountered higher ground.

This chapter will tell the story of the western tongue of ice that flowed out of the Coast Range of western British Columbia and turned south to push more than halfway through present-day Washington, finally stopping 160 miles south of the U.S.-Canada border, just south of Olympia. The next chapter will discuss the lobe of ice that pushed down between the Cascades and the Rockies, disrupted the channel of the Columbia River, formed a huge lake in western Montana, and finally released some of the largest floods known to have occurred on earth.

Continental Slope Exposed

The ice sheet in the west was as much as 90 miles wide. Rocky highlands that would form the Gulf Islands of British Columbia and the San Juan Islands of Washington disappeared beneath the advance. Today the islands' surfaces show parallel gouges left by rocks stuck to the bottom of the moving ice sheet. The ice cleaned out broken rock in faults created when pieces of former ocean bottom jammed together millions of years before the ice arrived. When the ice melted, it left the Gulf and San Juan Islands.

So much of the earth's moisture was locked up in the ice on land that sea level was more than 300 feet lower than today. Puget Sound, which now occupies the Puget lowland, did not exist. Instead, there was a wide valley between the Cascades and the Olympics with streams draining north. The streams skirted the Olympics and turned west to flow through a valley separating today's Olympic Peninsula and Vancouver Island. Much of the continental slope was then exposed land, and the Columbia River continued 20 or 30 miles beyond today's coastline before it emptied into the ocean.

The ice sheet split as it encountered the Olympic Mountains, and a western lobe turned toward the ocean. The westward tongue, the Juan de Fuca lobe, probably extended beyond Cape Flattery. It covered the northwest point of the Olympic Peninsula, extending as far south as the Soleduck and Quillayute Rivers. The section of ice that slid east of the Olympics, the Puget lobe, disrupted the drainage pattern in the broad valley between the Cascades and Olympics. A wall of ice several hundred feet high moved down

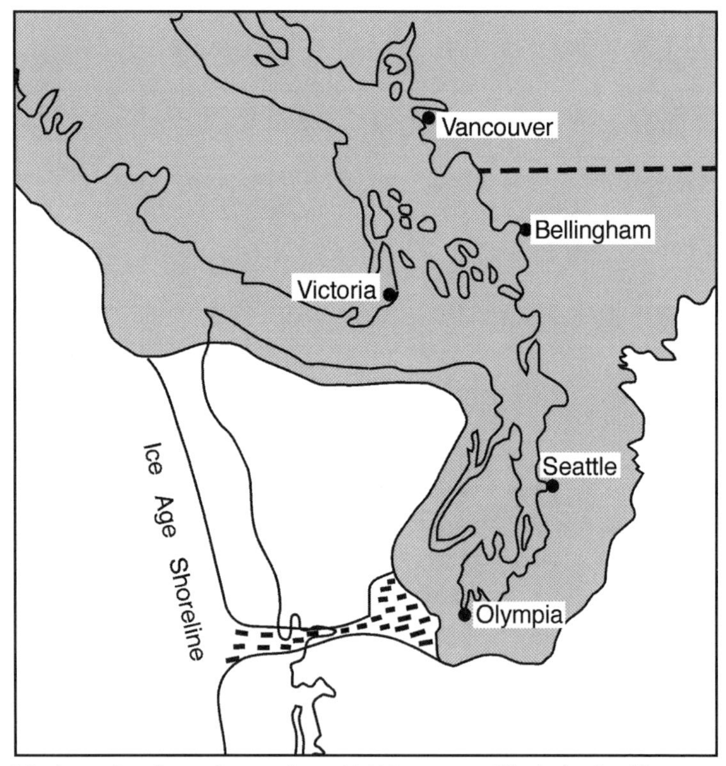

Maximum ice-sheet advance about 14,000 years ago fills the lowland between the Olympics and Cascades. (Short lines indicate area of drainage through today's Chehalis River gap and Grays Harbor.)

the valley, bulldozing ahead of itself a mound of rocks and dirt that blocked rivers and streams. Huge lakes formed in front of and to the sides of the advancing ice.

The Puget Lowland

The geological record is the same over much of the Puget lowland. The lowest, or oldest, deposits were left by the streams that once drained the valley. Above the stream deposits is a thick layer of glacial outwash, indicating a younger age. It began accumulating in the bottoms of lakes formed when the ice blocked the ancient drainage to the north. The earliest outwash deposits are made up of fine material, indicating they settled out of muddy water in standing lakes. Above that is coarser material that was carried away from the ice sheet by rivers seeking an outlet to the south.

Lakes covering much of the Puget lowland rose until they finally found an escape through the channel of the Chehalis River. During the 1,000 to 2,000 years that the ice lay on the Puget lowland, meltwater from the ice sheet and from streams pouring out of the Cascades flowed through the Chehalis River gap in the Coast Range, through Grays Harbor, and out over the exposed continental shelf to the ocean. That is why the Chehalis gap of today seems oversized for the river occupying it. It once carried huge volumes of meltwater.

The outwash accumulating ahead of the advancing ice sheet filled and leveled the Puget lowland. It is important in our story because it is the thickest layer of sediment left by the ice; it is the material from which the ice carved north-south hills and scooped out troughs. When the ice actually arrived it left a layer of till—blue-gray clay mixed with gravel—on top of the outwash. The bigger troughs, those filled today by the waters of Puget Sound, Hood Canal, and lakes Washington and Sammamish, probably were dug by full-size rivers running beneath the ice sheet.

The Maximum Advance

At its maximum, the ice was more than a mile thick at the Canadian border, half a mile thick at Seattle, and tapered to several hundred feet at the terminus south of Olympia. North of Glacier Peak in Washington's North Cascades and on into British Columbia, the mountains were largely buried by ice. It was a scene similar to the ice-capped mountainous parts of Greenland and Antarctica today. South of Glacier Peak, the ice-age Cascades probably resembled today's Alaska Range in south central Alaska, with its big alpine glaciers.

The ice sheet advanced an average of 300 feet a year—the length of a football field—at times faster, at times slower. It reached its maximum advance between the Cascades and Olympics around 14,000 years ago, stopping against a range of low hills about 15 miles south of Olympia. One thousand years earlier, those hills wouldn't have done more than slow the ice sheet's advance. But the ice sheet was dying. As the climate began to warm, the North Cascades and the mountains of British Columbia were no longer sending glaciers down into the valleys. It's interesting that the glaciers high in the mountains east of Seattle had begun retreating long before the main ice sheet ground to a stop. The alpine glaciers were smaller and able to react more quickly to climatic change. The big ice sheet was like a huge ship that plows ahead for hours even after its propellers stop.

The Ice Sheet Retreats

When the ice finally stopped moving and began to melt, its retreat was as chaotic as its advance. Meltwater again formed huge, temporary lakes that drained to the sea through the Chehalis gap. Lakes that had been bottled up in valleys in the Cascades and the Olympics began to drain. The climate warmed rapidly, perhaps more quickly than at the end of earlier ice ages. The ice sheet fell

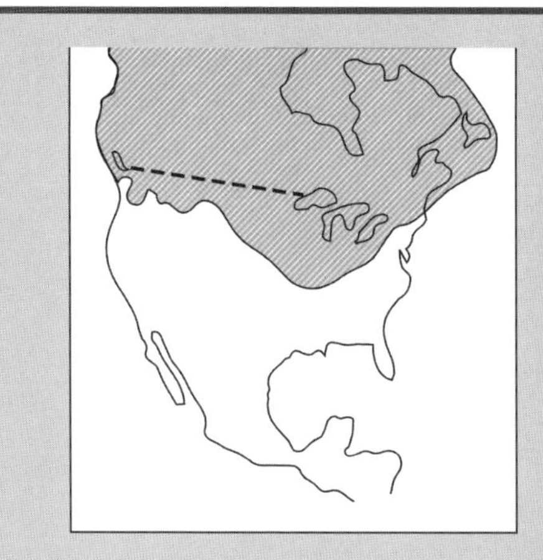

Earlier Ice Ages

The ice that covered northern portions of the Pacific Northwest until about 13,000 years ago was one of several ice sheets that together buried almost all of Canada, northern Europe, Scandinavia, the British Isles, and part of Russia. The Alps were too far south to be covered by the ice sheet, but they were mantled with huge glaciers, as were high mountains all over the world including those in the Southern Hemisphere.

An immense ice sheet that formed over Hudson Bay moved south, digging the Great Lakes and going on to cover a huge area of the Midwest. The Missouri and Ohio rivers mark the furthest extent of the ice sheet. Both rivers were forced into channels along the edges of the ice sheet and remained there when the ice retreated. The ice caps that cover Antarctica and Greenland today are modern reminders of the way much of the Northern Hemisphere appeared during the ice age.

Geologists can find clear evidence on land of several such ice ages during the past million years or so, including the most recent one. But a record preserved in deep-sea sediments suggests that between 1.5 million years ago and 10,000 years ago (the period known as the Pleistocene Epoch) there were as many as twenty ice advances. An important clue in the sediment is the variation in the ratio between two forms of oxygen with different molecular weights. The variations are believed to reflect changes in the volume of ice worldwide and, thus, to represent a record of ice ages.

apart so quickly that sometimes huge chunks of ice were left behind, buried under sediment. When the chunks melted—a process that could take a thousand years or more— they left kettle-shaped holes that geologists, fittingly, call "kettles."

Lake Washington and Seattle were clear of ice by 13,500 years ago. As the Puget and Juan de Fuca lobes retreated toward each other, seawater began trickling into the deepened Strait of Juan de Fuca, the Strait of Georgia, and Puget Sound. Water surrounded the highlands in the Strait of Georgia, creating the Gulf and San Juan islands. The flooding of lowlands was accelerated by a rise in sea level as vast volumes of ice around the Northern Hemisphere melted and drained into the ocean. At the same time, the surface of the earth that had been squashed down by the tremendous weight of ice was rebounding, a slow process continuing today.

The geological record for this period is complicated to sort out because sea level was rising rapidly at the same time as the earth's surface was rebounding. But by 10,000 years ago, the Ice Age was over in the Pacific Northwest. Mountain glaciers in British Columbia and Washington were no more extensive than they are today. Puget Sound was filled with salt water. The coastlines of British Columbia, Washington, and Oregon assumed their familiar outline as the rising ocean covered the continental shelf.

Lasting Impact of the Ice

The ice ages left a changed Pacific Northwest. Mountains, previously high but rounded, had been carved and gouged into spectacular peaks. Alpine glaciers flowing to the sea had left deep valleys, later flooded by the rising sea to create the scenic fjords of the British Columbia coast.

Reminders of the ice age include boulders uprooted in mountains, carried along with the ice, and then dumped when the ice melted. Composition of many of these boulders shows they originated in the mountains of British Columbia. Thousands of years after the ice dumped them, they still sit among houses in the Seattle area and farther south.

Along with the scenery, the ice left a sizable problem for today's residents of the Seattle area: unstable layers in the ice-carved

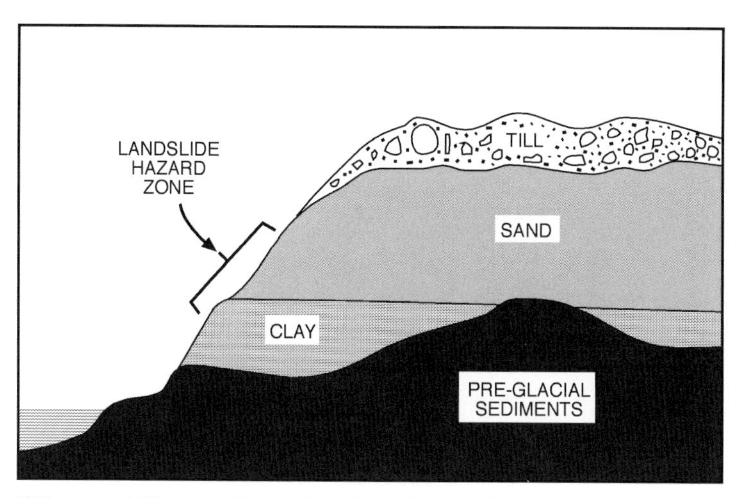

Slide-prone hills around Puget Sound are a legacy of the ice age.

hills. During a wet winter, water seeping into hillsides may accumulate at the boundary between layers, acting as a lubricant. If the slope is steep enough and the boundary slippery enough, an entire hillside may slide down, carrying houses and cars with it.

Even some of Seattle's traffic problems can be blamed on the ice age; the hills and valleys carved by the ice provide easy routes for north-south traffic but create obstacles for east-west commuters.

❖❖❖

15,000 years ago: ice plugs a river canyon and backs up a huge lake, repeatedly sending catastrophic floods hundreds of miles to the sea.

11

Glacial Lake Missoula Floods

The great ice sheet that lay over British Columbia at the end of the Ice Age was dying. But it still had enough energy for one last advance. Ice was encroaching on the northern edge of the basin that lay between the Cascade Range and the Rocky Mountains. But instead of finding relatively easy going in a wide valley, as happened in the Puget lowland, the ice encountered highlands in what is now northeast Washington and north Idaho. Tongues of ice probed ahead of the main ice sheet, entering low spots between highland ridges, sometimes filling entire valleys.

The advancing ice was setting the stage for a series of floods that would rip across the Columbia Basin and through the Cascades to the sea, some of the greatest floods known to have occurred on Earth. They were catastrophic, almost unimaginable, and humans probably have never witnessed anything like them. Even today the scars and torn-up landscape left by the floods constitute some of the most extraordinary scenery in eastern Washington and northern Oregon.

The floods were the last great shapers of the Pacific Northwest, capping a series of geological events that had begun more than one billion years before.

The Columbia Basin

Here is how the floods came about: the ice lobes moved along river valleys that, before the ice arrived, had drained toward an ancient basin between the Cascades and Rockies. As described earlier, the basin, once swampy, had been overrun by repeated flows of lava millions of years earlier. Under the accumulating weight, the basin sank even more, becoming saucer-shaped. Over millions of years, the bare rock had been covered over by wind-deposited soil to a depth of up to a hundred or more feet—the result of thousands upon thousands of dust storms. As the Idaho mountains had risen, the eastern edge of the basin had tilted up slightly, lowering the opposite side of the basin near present-day Pasco in south central Washington. The Columbia Basin, as it came to be called, was aligned to funnel the coming floodwaters across Washington.

Even when the great ice sheet lay just to the north, the Columbia Basin was not too much different from today. The climate was wetter and cooler, and the Cascades and Rockies were heavily glaciated. But the basin was free of snow in the summers and there was enough grassy and shrubby vegetation to support mammoth, mastodon, bison, caribou, and musk ox, which were preyed on by wolves and saber-toothed tigers. The Columbia River, its northern channel disrupted by encroaching ice and probably swollen with glacial meltwater and sediment, had spread out in several channels across the basin, all flowing toward the low spot, Pasco basin. There is no evidence that humans were in the area at the time of the great floods, although they may have witnessed later smaller floods as the ice age waned.

Glacial Lake Missoula

A set of circumstances had to be just right to cause such great floods. And Nature pulled it off repeatedly from about 15,000 to 13,000 years ago. First of all, one of the lobes probing ahead of

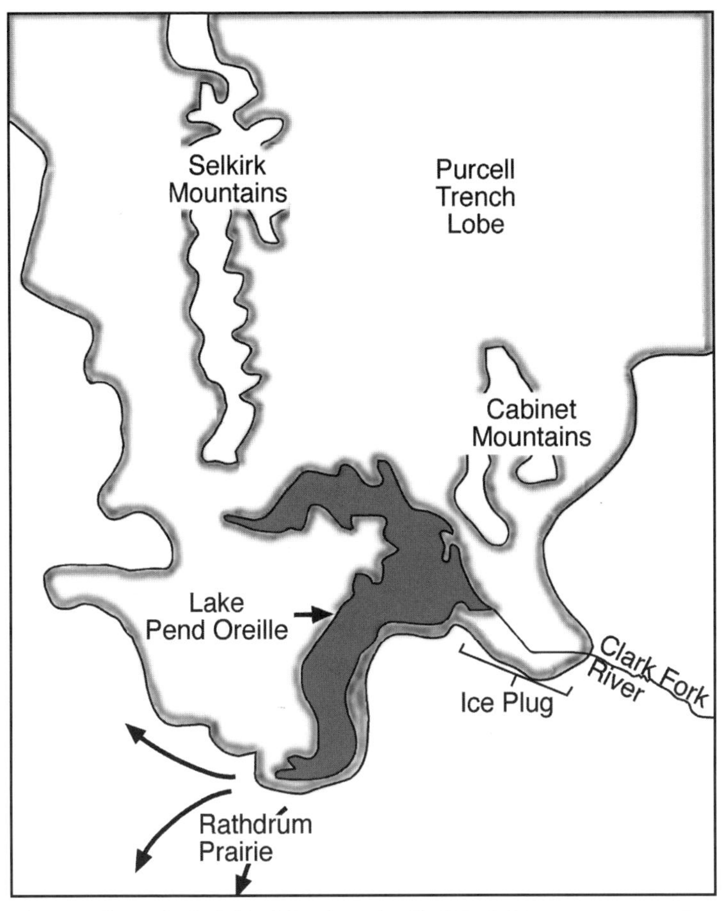

The Purcell trench glacial ice lobe plugs the Clark Fork River canyon, covering ancient Lake Pend Oreille in what is now Idaho. As a result, Glacial Lake Missoula was created to the east in the Clark Fork River drainage.

the ice sheet bulldozed its way through the Purcell trench, a valley that had been created just west of the Rocky Mountains millions of years earlier. As the Purcell trench lobe inched its way south, it crossed a river canyon and filled a lakebed dredged out by an earlier ice advance. The stream, which not only drained the lowland ice sheet but also glaciers in the Rockies, is known today as the Clark Fork River. It flows west out of Montana and empties into Lake Pend Oreille in northern Idaho.

As ice blocked the canyon, water began backing up into the valleys of northwest Montana, slowly filling a huge lake. The extinct lake, now known as Glacial Lake Missoula, had an estimated maximum volume of 530 cubic miles of water, about half that of today's Lake Michigan. Erosion marks left on mountainsides by the lake's varying levels show that water was as deep as 2,000 feet where ice plugged the canyon. The ice dam holding back Glacial Lake Missoula was nothing like today's concrete dams. Instead, it was a thick mound of ice that blocked about 30 miles of the canyon, extending over Lake Pend Oreille.

The Glacial Lake Missoula Floods

One would assume that eventually the backed-up lake would have risen high enough to flow over the ice plug or to find channels through the Bitterroot Mountains. But there is no evidence of that happening. Instead, the lake apparently burst down the river's old canyon.

Here is what some scientists believe happened, based on similar but much smaller outbursts from a glacial lake in Iceland: Glacial Lake Missoula would rise to a point where it began to float the edge of the ice. Water at the base of the ice plug, under pressure from almost half a mile of water above it, would wedge farther and farther beneath the ice until it opened one or more tunnels.

The drainage would quickly have become a torrent, probably reaching its maximum within a matter of hours. Although Glacial Lake Missoula lay at the edge of the continental ice sheet, its water was slightly warmer than freezing. This relative warmth

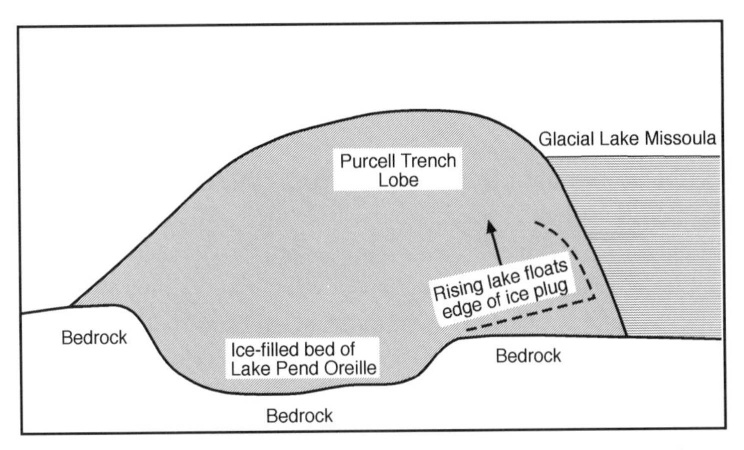

A simplified cross-section shows how rising water in the glacial lake may have floated the edge of the ice plug, allowing water to tunnel through the blockage.

would have quickly enlarged the tunnels. In some instances, the tunnels in the ice plug would have grown so rapidly that their roofs would have collapsed, opening a breach in the dam and allowing the entire lake to drain. Each flow probably ended as quickly as it started. As the lake emptied, the ice dam (without the water to support it) would collapse, sealing off the canyon and allowing the lake to begin rising again.

How many floods were there?

The great floods repeatedly roared across the Columbia Basin during the ice age's last gasps. How many times? Opinions vary from half a dozen to as many as 90. Although the number of floods is still controversial, a growing group of geologists now believes there were at least 30 to 40 catastrophic floods in the 2,000-year span, followed by smaller ones as the ice sheet shrank. Explaining how there was sufficient water to refill Glacial Lake Missoula many times doesn't seem to be a problem. The present flow of the Clark

Fork River would fill the lake in about 135 years. Adding ice-age meltwater loaded with silt and debris that poured off the ice sheet and mountain glaciers cuts the estimated refill time to 60 or so years.

Each flood emerged from the edge of the ice lobe and roared across the Columbia Basin—the huge tilted saucer—toward the low point in the Pasco basin. The outbursts were similar to the floods of molten basaltic lava that had overrun the Columbia Basin millions of years earlier. They, too, had originated at the eastern edge of the basin and flowed toward the sea. But whereas the basalt had filled canyons, leaving a scorched and level plain, these floods

The Great Controversy

There is no doubt today that great floods ripped across the Columbia Basin late in the ice age. But from the time it was first suggested, more than 30 years went by before the geological community accepted the idea.

J Harlen Bretz, a University of Chicago geologist who had once taught high school in Seattle, proposed as early as 1910 that the Columbia Basin had been swept by a great ice-age flood that sculpted the channeled scablands. Many geologists of the time scoffed at the preposterous idea of a monstrous, "catastrophic" flood.

Bretz's problem was that he couldn't explain where such quantities of water had come from. It wasn't until 1940 that another geologist reported evidence that a huge glacial lake had backed up behind an ice plug in the Clark Fork River—and the lake had drained suddenly!

Still, geologists were slow to connect Glacial Lake Missoula with Bretz's flood idea. Luckily, Bretz lived into his nineties, long enough to see his ideas validated. Field trips that allowed unbelievers to actually see the geological evidence in the scablands turned the trick. After a 1965 field trip that included several hard-core opponents of the flood idea, Bretz received a telegram admitting, "We are now all catastrophists." In 1979 Bretz was awarded the Penrose Medal, the nation's highest geological honor.

of icy water ripped up the rock, eventually leaving gaping dry channels and exposed basalt that puzzled the first geologists to study them. Early settlers called the scarred areas the scablands. As the flood idea was accepted, the name became the channeled scablands.

From Breakout to the Sea

Each flood was different in its details, but let's trace a typical flood from the great breakout in northern Idaho to the sea. As water burst from the canyon it turned south to follow a broad valley, today's Rathdrum Prairie. Imagine the scene if we could have been standing in a safe place to watch: it's possible that waves of pressurized air pushing ahead of the flood toppled trees before the water arrived. Then a torrent of water loaded with trees, boulders, and icebergs rushed past, the ground shaking from the thundering weight of the water.

Try to visualize this: at the flood's maximum, more than 9 cubic miles of water roared past a given point in the valley every hour! The flow was equivalent to 10 times the combined flow of all the rivers in the world today, or 1,000 times the flow of today's Columbia River. The slope for the first 70 miles or so was the steepest of the flood's entire run to the sea, a drop of more than 2,000 feet from the crest of Glacial Lake Missoula. That gradient would have sent the torrent rushing along the valley at more than 50 miles an hour. The flood entered, filled, and overflowed the Spokane River valley and turned west as if a giant nozzle were aiming a high-pressure stream at Spokane. The flood was as much as 500 feet deep as it flowed over the future site of Spokane, slopped over low hills, and spilled into the Columbia Basin.

The torrent spread out over the basin, following existing drainages and scooping out new ones. There were three huge rivers simultaneously roaring across the basin, interconnected to make a braided pattern visible today in aerial photographs. Geologists

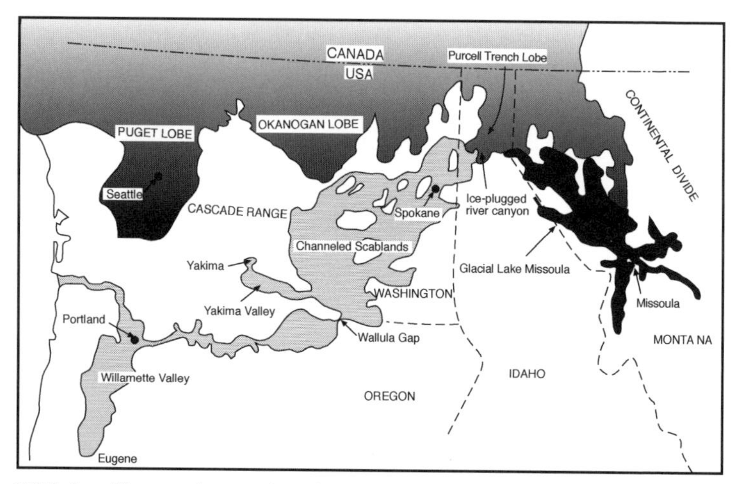

While ice still covered parts of northern Washington and Idaho (shown by graded shading), gigantic floods burst from Glacial Lake Missoula (shown as black area) in Montana and roared to the sea. The lighter shaded pattern shows the area carved by the floods.

estimate that nearly 50 cubic miles of soil and basalt had been excavated from the Columbia Basin by the time the great floods ended. They left dry falls, deep canyons, exposed rock, and braided channels that today are windswept and bone dry.

Wallula Gap

The channels all eventually came together at the basin's low spot near Pasco. And there, a few hours and about 180 miles from the ruptured ice dam, the torrent encountered a bottleneck: the gap where the Columbia River had maintained its channel through the Horse Heaven Hills. The one-mile-wide channel, today's Wallula Gap, couldn't accommodate the vast flow. Water backed up more than 100 miles into the Snake River valley, creating a temporary lake almost 600 feet deep at the point where the

Today's Columbia River flows through Wallula Gap, a bottleneck in the great ice-age floods.

Clearwater River flows into the Snake River at the present site of Lewiston, Idaho.

The vast backed-up waters would have been studded with icebergs, many of them carrying huge boulders pried from mountainsides by moving glaciers. Wallula Gap's layered basalt walls are 1,000 feet high, yet the flood overflowed the hills on either side. U.S. Highway 730, which today runs along the river through the gap, would have been under 1,000 feet of water.

Water poured through Wallula Gap at the stunning rate of more than 40 cubic miles a day. It would have taken about two weeks for the flow to drain through the gap, long enough for the silt-laden water to drop a layer of sediment. Layers laid down by the temporary lakes formed after each flood were to serve as markers for geologists thousands of years later.

The rushing water created another huge temporary lake as deep as 1,000 feet, covering the Umatilla, John Day, and The Dalles basins on the Oregon side of the river. Hillsides bordering the basins show the scars of high channels where floodwater actually topped the hills and fell back into the river valleys. This backed-up pool was from 10 to 30 miles wide and extended 100 miles downstream from Wallula Gap to the present site of The Dalles Dam, where the river narrowed again.

The Columbia Gorge

Below The Dalles the flood began funneling through the channel cut by the Columbia as the Cascade Mountains were uplifted. Some geologists believe the canyon was a typical narrow-sided river canyon and that the great floods widened it and changed its V-shape to a flat-bottomed U-shape. As the flood cut back the canyon walls, streams that formerly had tumbled down steep slopes to the river were left hanging, creating the beautiful waterfalls in today's Columbia Gorge.

The flood burst from the gorge (possibly topping Crown Point where the familiar Vista House is located today) and spread over the Portland-Vancouver plain. A

The Willamette Meteorite

One of the most interesting pieces of glacial debris left by the Glacial Lake Missoula floods is a 15-ton meteorite discovered in 1902 on a hill about ten miles south of Portland. The meteorite's resting place lacks any evidence of a violent impact. So it appears that the meteorite, the largest ever discovered in the United States, must have fallen onto, or perhaps been picked up by, the moving ice sheet in Montana, northern Idaho, or Canada.

Hot enough on impact to sizzle ice into steam, the big meteorite was frozen into the ice and carried slowly toward Glacial Lake Missoula. When the lake was breached, the meteorite, stuck in what was now an iceberg, rode in the flood across Washington State and through the Cascades. Pushed along by a wall of backwater surging up the Willamette Valley, the iceberg grounded on a hill above the Tualitin Valley and melted. There the "passenger" meteorite remained.

The Willamette meteorite is in the American Museum of Natural History in New York City.

narrowing of the river channel downstream from there backed up water in the Willamette Valley as far south as Eugene, Oregon. The present site of Portland was under 400 feet of water; only Rocky Butte and Mount Tabor remained above water.

After squeezing through the narrows near what is now Kalama, Washington, the torrent's height dropped rapidly as it approached the ocean. The crest of the floods was near present-day sea level as it reached Astoria. But, with the ice-age sea level still lower than today, the flood followed the Columbia's channel across the continental shelf before it reached the ocean.

It is impressive that one or more of the ice-age floods, even after rushing hundreds of miles to the sea from the ice breakout in northern Idaho, still had the power to trigger ocean-bottom turbidity currents loaded with sand and gravel that plowed across hundreds of miles of seafloor.

How did gravel get carried out to sea?

Oregon State University oceanographers were briefly puzzled by an ocean-bottom discovery of the late 1960s. Drilling into the Cascadia channel from a research ship, they found a gravel layer seven feet thick, containing rocks as big as 1.5 inches in diameter. Mineral composition of the rocks showed that they came from the Columbia River's drainage area in the Rocky Mountains. Adding to the mystery, the gravel layer lay under 10,000 feet of water in the Cascadia basin, 450 miles out from the coast. The Columbia, mighty as it is, couldn't carry rocks of that size hundreds of miles out to sea.

Fossils of tiny marine animals in the gravel showed the layers had been deposited late in the ice age. The oceanographers quickly realized the gravel layers must be a result of the catastrophic floods. They calculated that an ocean bottom current must have been moving at least 13 miles an hour to move rocks of that size. That may not sound very fast compared to automobile speeds, but it's impressive that a turbidity current almost as thick and heavy as concrete once plowed hundreds of miles across the ocean floor at speeds exceeding 13 miles per hour.

Evidence of Ancient Floods

Grand Coulee

Of all the changes wrought in the landscape by the Glacial Lake Missoula floods, the Grand Coulee in central Washington is probably the most spectacular: 50 miles long, 1 to 6 miles wide, with walls up to 900 feet high. Grand Coulee's story began long before the great floods. An earlier advance of the ice sheet had blocked a northern loop of the Columbia River's course and diverted the swollen stream southwest along a gentle depression. The river carved a modest channel, which it abandoned when an ice retreat allowed it to return to its former channel.

The temporary river channel was in place—perhaps dry or perhaps again serving as the Columbia's course during another ice advance—when an outburst flood from northern Idaho sent a torrent through the Spokane River valley along the southern edge of the ice sheet. The flood encountered the ice and, like the Columbia before it, swerved into the temporary bed of the river.

What happened next almost defies description. The flood quickly filled and overflowed the existing channel. Water hundreds of feet deep rushed across nearby hills and poured into valleys. In the channel itself, the tremendous pressure of rushing water ripped into the basalt, tearing out house-size boulders and sweeping them downstream, and deepening and widening the canyon.

A cataract developed at the canyon's lower end where water fell into a depression. Erosion was rapid at the lip of the cataract and the waterfall worked its way upstream for miles, probably in a matter of days. But the flood was so great—15 miles wide and

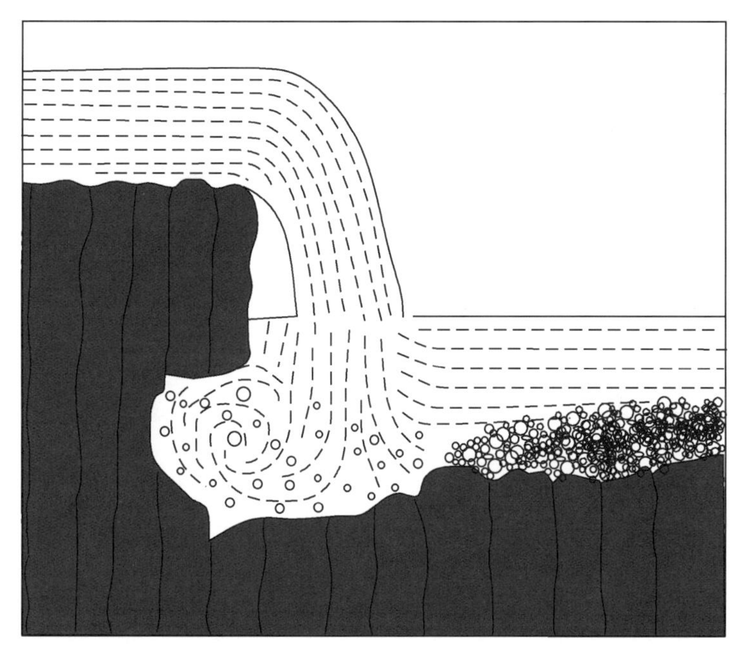

Torrential water pouring over a ledge undercuts and collapses the ledge, moving the waterfall upstream. Moses and Grand coulees and the Palouse River canyon were carved rapidly in this way by ice-age floods.

200 feet deep—that there wouldn't have been an actual waterfall during the flood itself. Instead, the falls would have been eroded underwater; at the surface there would have been just a drop in the rushing water.

Only when Glacial Lake Missoula emptied and the flood stopped did the world-famous Dry Falls emerge to mark the place where the canyon-excavating cataract stopped. Visitors to the Dry Falls State Park viewpoint today can see only a part of the ancient falls—cliffs 350 feet high (higher than Niagara Falls) that extend for 3 miles. It is awe-inspiring to remember that not once, but many times, rushing water 200 feet deep poured over a 15 mile section of those cliffs.

Moses Coulee

About ten miles to the west of Grand Coulee is another yawning canyon carved from solid rock by rushing floodwaters. A couple of centuries or so before the Grand Coulee episode, a flood from a Glacial Lake Missoula breakout created this smaller, but still impressive, coulee. A 40-mile-long gash, it runs parallel to the larger canyon, in alignment with the course of the great floods.

The Moses Coulee flood emerged from the Spokane River valley and then took a shortcut across country, dropping over steep cliffs into the Columbia River a dozen miles downstream from present-day Wenatchee. Under the onslaught of the flood, the lip of the cataract rapidly eroded upstream, leaving a gaping coulee behind it. When the flood ended, it left Moses Coulee's dry falls, with the deepest part of the coulee—walls up to 1,000 feet high—located downstream. Above the dry falls, where a flood torrent 300 feet deep and 5 miles wide raced toward the drop, Moses Coulee is shallower, partly filled by sediment dropped as the flood died.

The Palouse Canyon

Before the floods, the Palouse River flowed generally southwest across the basin to join the Columbia upstream from Pasco. That changed when a flood filled the valley of the Palouse River, slopped over hills to the south, and cut a new, shorter channel to the Snake River. When the torrent fell into the Snake, it began cutting a canyon upstream, much as happened at Moses and Grand coulees. When the flood stopped suddenly, Palouse Falls emerged several miles upstream from the Snake. The Palouse remained in its new channel, leaving a real waterfall, Palouse Falls, today. At 198 feet, it is higher than Niagara Falls, although much narrower.

The Touchet Beds

When the floods slowed at Wallula Gap to form the huge temporary lake, water surged upstream in the rivers that had drained into the basin. The powerful surge briefly flooded the Walla Walla and Yakima River valleys and left one of the most intriguing clues of the series of floods. When the surging water from each flood stopped and was still for days or weeks before draining back toward Wallula Gap, a layer of sediment settled to the bottom of the Walla Walla and Yakima Valleys. Successive floods left their own

Flood-deposited layers in the Walla Walla Valley are believed to be evidence of great ice-age floods. *Steve Reidel photo*

deposits until layers were stacked like pancakes in the valley bottoms. Erosion has exposed the layers in several places.

The deposits are known as the Touchet beds for a river in the Walla Walla Valley. But similar beds in other valleys tell the same story: water surging upstream, pausing long enough to drop sediment, and then draining away. Several riverbank bluffs in those valleys reveal more than 30 layers. Each layer has coarse material at the bottom, and graded finer sand and silt at the top, exactly what would be expected if flood water slowed and stopped: coarse material would settle first, followed by finer sediment.

Some geologists question whether each layer represents an individual flood or whether several layers could represent separate surges of one flood. Still, some layers show evidence of soil formation at the top, indicating that many years had passed before the next layer was deposited.

Huge icebergs floated in the temporary lakes that backed up above Wallula Gap and farther downstream along the Columbia

Another Great Flood

The cool, wet climate of the ice age created other lakes in the West, scattered across Nevada and Utah in today's bone-dry Great Basin. The biggest of these was Lake Bonneville, which covered much of western and northern Utah. The Great Salt Lake is its tiny remnant.

Fed by melting snow and ice in the Sierra Nevada to the west and the Wasatch Range to the east, Lake Bonneville rose to a depth of almost 1,000 feet. (By contrast, today's Great Salt Lake averages about 10 feet deep.) The sites of today's Salt Lake City and neighboring cities were deep underwater.

About 15,000 years ago, during what must have been a particularly wet season, the lake began overflowing to the north. The overflow stream quickly cut down through easily eroded rock to deepen and widen the channel. Within weeks, at least a third of huge Lake Bonneville had spilled north into the Snake River valley of Idaho and roared west across lava plains. The flood greatly enlarged the canyon, explaining the seemingly oversized channel for much of the length of today's Snake River. Travelers on Interstate 84 near the Utah-Idaho border can travel the flood route through Red Rock Canyon, the point where Lake Bonneville broke through its banks.

channel. Some had boulders embedded in them. As the lakes drained, some of the icebergs grounded and were left behind. As they melted, the embedded boulders remained. Many of them still stand on hillsides above valleys once flooded by ice-age lakes, silent indicators that the water once was at least that high.

Long Beach Peninsula

As the violence of the great floods faded, gentler natural processes utilized the vast quantities of material dumped by the floods to begin putting finishing touches on the Pacific Northwest as we know it today. For thousands of years after the ice sheets melted, the Columbia River carried huge quantities of sand and silt from glaciated or flooded areas to the ocean. Waves and currents carried most of the sand north from the river's mouth and deposited it to build the Long Beach Peninsula, a 27-mile-long strip of sand that shields Willapa Bay from the open sea.

Dredging of the river's mouth and construction of jetties may have spurred natural accretion; for whatever reason, the peninsula has widened as much as a half mile in the past century. As recently as the late 1970s, the Long Beach Peninsula was still growing. But by 1990, accretion had stopped. Erosion, which began in the 1980s at the peninsula's extreme south end, may be next. Many geologists attribute the change to the damming of the Columbia, which has slowed the current so that sediment now gathers in reservoirs behind dams and has largely eliminated spring floods which formerly flushed huge amounts of sand into the ocean.

Windblown Soil

In channels all across the Columbia Basin, floods scoured away wind-deposited silt that had accumulated over thousands of years. Where hills stood above the flood, the soil was spared, forming streamlined islands rising above the channels. Many of eastern

Washington's wheat fields are on hills that escaped flood erosion. Farmers grow wheat on these "islands" that overlook dry, barren flood channels where only sagebrush grows. Much of the flooded area is still so jumbled with exposed rock and dry channels that the drainage of surface water is chaotic; nature has not had time to establish throughgoing streams.

As the Columbia moved the material destined for the Long Beach Peninsula, wind began building another feature in the Pacific Northwest—the Palouse soil that supports one of the world's most important breadbaskets. As the backed-up lake above Wallula Gap emptied after each of the catastrophic floods, it left expanses of fine, dry material that swirled up into dust storms. The prevailing winds thousands of years ago must have been toward the northeast, the same as today. The floods left the heaviest deposits of sediments in the Pasco basin, so it is no coincidence that today the area with the thickest topsoil is directly downwind from the basin. In the Pullman-Colfax area of southeast Washington, the windblown soil can be as much as 200 feet deep. Scientists say the winds we experience today, if continued over thousands and thousands of years, can easily account for the thick Palouse soil. In fact, soil-building is continuing today. Residents of the Palouse country sometimes see it happening when falling snow appears dirty, carrying new soil to the hills. But unfortunately, the heavily cultivated Palouse soil is eroding into rivers much faster than nature replenishes it.

The scenic Long Beach Peninsula, the rich Palouse soil, the scarred channeled scablands, the river canyons, and the gaping coulees are all legacies of catastrophic floods of the late ice age.

❖❖❖

The 21st century and beyond: the possibility of great earthquakes and more volcanic activity.

12
Not If, But When

Hardly anyone was thinking seriously about the possibility of really big earthquakes in the Pacific Northwest back in the 1970s when oceanographers drilled into Cascadia channel, the long, winding canyon on the ocean floor off the Oregon-Washington coast. Evidence brought up by the drill was pretty much as expected: the channel had been repeatedly deluged by seafloor avalanches. The sediment was ice-age debris, eroded from the scarred land during the past 10,000 years and dumped offshore by the Columbia River.

Answers from Ash

Near the bottom of the channel deposits was a familiar and useful marker, a layer of ash blasted from the Mazama volcano about 6,900 years ago and carried to the sea by the Columbia River. The marker enabled scientists to date the sediment layers. As on land, layers below the Mazama ash were older than 6,900 years; those above, younger.

At the time, oceanographers assumed that the undersea avalanches had been caused by piles of sediment becoming overly steep, causing them to collapse and rush down the continental slope to deep water. But ideas had changed by the early 1990s. Evidence was accumulating that earthquakes more powerful than any recorded in the Pacific Northwest had occurred here in the distant past, long before written records began for this part of the country less than two centuries ago.

The new evidence of ancient earthquakes prompted another look at the decades-old records of the Cascadia channel drilling. Again, the Mazama ash layer was the key. It enabled scientists not only to date but to match up individual deposits at widely separated places along the channel, a little like matching growth rings in trees at different locations.

This second look revealed an intriguing fact: thirteen times since the Mazama ash marker, undersea avalanches from widely scattered places along the coast had been triggered simultaneously. This was not the expected pattern of a pile of sediment at one location suddenly failing and sluicing down the continental slope, and then perhaps centuries later the same thing happening at another place a hundred miles away. Instead, on thirteen different occasions these piles of glacial debris apparently let go simultaneously up and down the coast and, following different paths, fed into the Cascadia channel.

What triggered these undersea avalanches?

The simplest explanation for the simultaneous avalanches seemed to be surprisingly powerful earthquakes. Based on the thickness of layers of open-sea material that had drifted to the bottom and accumulated between avalanche deposits, scientists estimated that the great undersea avalanches had occurred fairly regularly, averaging about every 600 years—the most recent about 300 years ago.

And here the plot thickens. Geologists working in the muck of coastal estuaries had begun finding evidence in the late 1980s that the outer coast had repeatedly dropped abruptly in the past several thousand years, drowning vegetation and burying it in mud. The most recent drop of the outer coast? About 300 years ago.

The coastal clues, first noted in southwestWashington and later up and down the Washington and Oregon coasts, were exactly what would be expected if the Juan de Fuca plate, stuck tight beneath the edge of the continent, had suddenly broken loose and jumped forward.

By the 1990s, geologists, construction engineers, and many in the public realized we live in an area that has been shaken by great earthquakes and they are likely to occur again. The earth's stresses and strains that built the Pacific Northwest over millions of years are still at work.

The San Andreas Fault

Although the North American and Farallon plates have been colliding for hundreds of millions of years, a change in the collision zone created stresses still exerted on the Pacific Northwest today. The Farallon plate once underlay much of the eastern Pacific Ocean. Its eastward creep pushed it against and beneath the entire west coast of North America which was creeping west itself. About 29 million years ago, however, the continent approached and actually overrode part of the undersea spreading ridge that was creating the Farallon plate. This change occurred near the latitude of today's Southern California; it cut the Farallon plate into two pieces, both of which continued creeping toward the continent. The northern portion of the old Farallon plate is what we know as the Juan de Fuca plate, and it is still thrusting beneath the Pacific Northwest. The surviving portion to the south, the Cocos plate, dives beneath Mexico's west coast.

Between the surviving parts of the old Farallon plate, the continent and the Pacific plate were in direct contact. This created a grinding, restless fracture in the earth's crust where the continent pushed toward the west and the Pacific plate pulled toward the northwest; the two sides slide past each other more or less horizontally. Occasionally, as Californians know all too well, the creeping plates stick together, build up pressure, and then break loose in an earthquake. Today we know this long West Coast crack as the San Andreas fault.

As North America advanced, the San Andreas fault grew longer, eventually reaching a length of 800 miles. The tugging of the Pacific plate opened the Gulf of California and began pulling at California's coast. Today the Pacific plate is carrying along with it Mexico's Baja California and California's coastal strip south of the San Francisco Bay area. The San Andreas cuts through coastal California to a point north of San Francisco, where it heads out to sea. The coastal strip, including Los Angeles, is creeping northwest as the Pacific plate scrapes along the advancing edge of the continent. If the movement continues in the same direction, Los Angeles will be at the present latitude of Seattle in about 50 million years, moving at the stately speed of 1 1/2 to 2 inches per year.

Interestingly, the undersea portion of the San Andreas fault turns toward shore north of Vancouver Island, where it runs just offshore from the Queen Charlotte Islands. This extension of the San Andreas fault is known as the Queen Charlotte fault; farther north, off the coast of Alaska, it is known as the Fairweather fault. Whatever the name, it is the same fault—a zone of direct contact where the enormous Pacific plate muscles past the smaller North American plate.

The Dying Juan de Fuca Plate

Although the Pacific Northwest rides on the North American plate, we are buffered from its collision with the massive Pacific plate by the Juan de Fuca plate. Although the Juan de Fuca plate (equivalent in size to the combined area of Washington and Oregon) may seem big to us, it is tiny compared to the giants jostling it on both sides. Remember that the North American plate includes half of the Atlantic Ocean floor and all of North America; and that the Pacific plate, which underlies most of the Pacific Ocean, is the planet's biggest plate.

Probably due in part to the jostling, the Juan de Fuca plate appears to be dying, although there's no need to plan a funeral for many millions of years. Its movement toward the continent has slowed. Large chunks have broken off both ends and appear to be moving independently.

Off the northern end of Vancouver Island, the Explorer plate, once firmly attached to the Juan de Fuca plate, broke off 3 million to 4 million years ago and has developed its own motion. In fact, at least part of the Explorer plate may have been captured by the North American plate. If so, that would mean it no longer is thrusting beneath the continent, which would significantly reduce the danger of great earthquakes in the northern part of Vancouver Island.

The south end of the Juan de Fuca plate has also broken away to form what geologists call the Gorda plate off Northern California. It, too, appears to be moving independently of the Juan de Fuca plate and to have ceased moving toward the continent. There is speculation that the Gorda may have been captured by the Pacific plate and is actually pulling away from the continent.

But even in its dying stages, the Juan de Fuca plate still squeezes the Pacific Northwest hard enough that the results have actually been measured. Precise distances between markers on mountain tops in the Olympic Mountains of Washington have been measured with laser instruments and then remeasured 8 years later. Over a distance of 16 miles there was a tiny but significant shrinkage. A joint U.S.-Canadian project found a similar shortening across the Strait of Juan de Fuca. Preliminary results indicate that Neah Bay, at the northwest tip of Washington's Olympic Peninsula, is moving northeast at the rate of almost half an inch a year.

In all these cases, the direction of shortening or movement closely matches the angle with which the Juan de Fuca plate collides with the continent, north 50 degrees east. A logical explanation for this compression is that the subducting plate is stuck and building up pressure. The concern is that if it should break loose, it could cause a major earthquake. Unfortunately, it is likely that such great quakes have happened before.

The Big Squeeze

There is other evidence that western Oregon and Washington are being squeezed. Geodetic measurements repeated over the years reveal that the Coast ranges of Oregon and Washington are being tilted to the east. Tidal studies have confirmed that the outer coast tilts, too, causing something like a teeter-totter effect. Measuring stations at seaside towns such as Astoria, Oregon, and Neah Bay, Washington, indicate that along the outer coast of Oregon and Washington, land is rising. At the other end of the teeter-totter, east of the Coast Range (around Seattle, the San Juan Islands, and Vancouver, B.C.), land is sinking.

On the ocean end of the teeter-totter are estuaries where researchers skim along narrow tidal channels in canoes and wade through deep mud in hip boots. Banks cut by the back-and-forth tidal currents have exposed a record of plant life over the past several thousand years. In that record, geologists have found layers of ancient soil surfaces with remains of dry-land plants and trees that were suddenly covered with mud while the vegetation was still alive, indicating the outer coast had dropped abruptly as much as 6 feet. Atop some of the layers of buried vegetation remnants, researchers found a thin layer of sand similar to those left by tsunamis elsewhere in the world.

Scientists interpreted the presence of the sand as evidence that a sudden change in the ocean bottom offshore had sent a huge wave crashing against the coastline. The sequence in the estuary banks was repetitive—dry-land plants and trees covered by mud with a layer of sand, overlain by another layer of dry-land vegetation again covered by mud. Whatever the cause, it had happened repeatedly over thousands of years.

The shortening, the tilting, and the teeter-totter effect of the outer coast all are compatible with a big squeeze being exerted on the Pacific Northwest. Unfortunately, it indicates the subducting plate is locked and building up pressure. If the Juan de Fuca plate were diving smoothly and steadily beneath the edge of the continent, it wouldn't affect the coastline very much. But if it got stuck, increasing pressure would tend to push up the outer coast very gradually over hundreds or thousands of years. Eventually pressure would build until the stuck portion of the colliding plates broke free and the diving plate jerked forward and down. With the pressure released, the coastal strip would drop suddenly, flooding vegetation under saltwater. The vegetation would die and be covered with mud as tides washed in and out. And then as the plates locked together again the relentless process would resume with a gradual ramping up of the coastal strip.

Deadly Subduction Quakes

Seismologists estimate that a stuck plate breaking loose could cause a subduction earthquake of magnitude 8 or 9, stronger than any recorded before in the Pacific Northwest. The most powerful earthquake recorded anywhere in the world since modern instruments have been available was a subduction quake off the west coast of Chile in 1960. A stuck plate broke loose and jerked ahead, creating an earthquake that registered an amazing 9.5 on the Richter scale. This event occurred in an area geologically similar to the Cascadia subduction zone off the coast of the Pacific Northwest.

Despite increasing evidence of true subduction earthquakes in the Pacific Northwest hundreds or thousands of years ago, there have been none in the almost two centuries that written records have been kept. The Puget Sound area was rocked by a magnitude 7.1 earthquake in 1949, a magnitude 6.5 quake in 1965, and a magnitude 6.8 quake in 2001. All were identified as intraplate, occurring within the subducting plate rather than being caused by movement of the plate itself. There is every reason to expect that intraplate quakes of similar magnitude will strike again.

A great subduction quake in the Pacific Northwest would be in a class by itself. The threat of widespread destruction from a sudden drop in the ground surface and the impact of a tsunami is greatest along the sparsely populated outer coast. But even farther inland in Portland, Seattle, and Vancouver, British Columbia, the strong and prolonged shaking from such a quake would undoubtedly cause loss of life and widespread damage.

Seattle faces a threat much closer to home than the intraplate quakes of the greater Puget Sound area or the subduction quakes of the outer coast. A severe earthquake occurred about 1,100 years ago, literally beneath the foundations of today's city. The ancient quake was shallow and local, not distant and deep as past quakes must have been. A repeat could be devastating.

A Curious Jog in the Bedrock

There had been hints for a long time that something major had happened beneath Seattle—sometime. Geologists knew of a curious jog in the bedrock along a line running generally northwest-southeast beneath Puget Sound and crossing beneath South Seattle. South of that line, bedrock is close to the surface; to the north, bedrock is as much as 3,000 feet below the surface, buried by debris left by the glaciers. By the late 1970s, geologists were beginning to assume the jog in the bedrock was a fault along which rock had once moved, although there was little evidence of how old the fault was or whether it might still be active.

Geologists' uncertainty about past activity along the fault changed in the 1990s with the coming together of an extraordinary collection of research results. Scientists extracted information from ancient landslides, tree rings, sheets of sand sandwiched between mud layers, a part of an island that had abruptly risen 20 feet, and trees that were killed in avalanches. These bits of information were compelling because they clustered around the Puget Sound area and seemed to have occurred about 1,100 years ago, strongly suggesting they had been triggered by the same event. In geological terms, 1,100 years ago is like the day before yesterday.

Suspect Number One for this devastation was, of course, a large earthquake. The pieces of evidence were all consistent with a thrust earthquake in which rock is being compressed and accumulating strain, until it suddenly breaks and one side pops up over the other. The buried jog in the bedrock, the assumed fault, lies at a 90-degree angle to the northeast-directed force exerted by the Juan de Fuca plate—exactly as one would expect rock to break under unrelenting pressure.

The realization that such a catastrophic event probably did occur was both surprising and sobering. Although this thrust earthquake would not have carried the force of a great subduction earthquake or even a deep intraplate quake, it was so shallow that the effects must have been horrendous at the future site of Seattle. A repeat could mean sudden uplift or subsidence of land in the city, extremely strong shaking, and the creation of a tsunami racing in Puget Sound. By this time the "assumed fault" of the 1970s had acquired a new name: the Seattle fault.

Evidence of the Seattle Fault

It's usually difficult to find evidence of thousand-year-old land movements, particularly in the wet Puget Sound area where erosion and vegetation hide clues. But thanks to an accident of nature that kept the evidence exposed, geologists found a spot on Bainbridge Island in Puget Sound where the beach had jumped as much as 20 feet. The change had been so rapid that there were no intermediate shorelines, as would be seen if the rise had been gradual. Studies of ancient vegetation showed that tide flats had suddenly changed to freshwater swamps and meadows without the usual intermediate step of tidal marshes. Radiocarbon dating of the killed vegetation yielded an age consistent with an event occurring about 1,100 years ago. The uplifted shoreline is on a point just across the sound from Seattle. It is on the south side of the Seattle fault, on the "high side" of the jog in bedrock.

If a large portion of the bottom of Puget Sound suddenly rose 20 or so feet, it would be expected to send a large wave sloshing away from the uplifted bottom. This is what seems to have happened when the south side of the Seattle fault jumped skyward. The telltale clue is a layer of sand found in two places along Puget Sound. The bed of sand, several inches thick, is similar to those left after tsunamis have washed ashore elsewhere in the world.

One of these sites is a beach deposit on Seattle's West Point and the other is off the south end of Whidbey Island—both of them on the north, or "low side," of the Seattle fault. Bits of wood and vegetation in the sand layers were found to date back to the time of the assumed quake on the Seattle fault. The level of the land at West Point underwent a sudden drop of approximately 3 feet around the time the sand layer was deposited. Such a subsidence north of the Seattle fault would be expected as a result of a large quake along the fault.

Clues from Sunken Forests

More evidence that the Seattle area was shaken severely about 1,100 years ago was locked up in several strange sunken forests in Lake Washington along Seattle's eastern edge. People had known for almost a century that the trees were there, many still standing erect on the lake bottom. Some that were deemed navigation hazards were trimmed off in 1919. But as interest grew in the possibility that large earthquakes occurred in the geological past, scientists took another look. Radiocarbon dating of wood taken from trees in three widely separated slides in Lake Washington showed that the trees had all died about 1,100 years ago. Despite heavy vegetation, it is still possible to identify scars left on the land when forested hillsides slumped into the lake, apparently with trees still standing. The proposed cause: a severe earthquake.

In another elegant touch, scientists took growth rings from seven submerged trees in Lake Washington and cross-dated them with rings in a log found in the tsunami deposit at West Point, about 10 miles away. It turned out that the log at West Point and the trees in Lake Washington's sunken forest had all died in the same season of the same year.

A final clue was drawn from the quiet mountain lakes in the Olympics west of Seattle. Here rock avalanches had thundered down slopes, dammed creeks, and formed lakes that flooded and killed many trees. Comparing growth rings showed that trees in three of the lakes had died about the same time, raising the possibility that an earthquake had shaken loose several avalanches simultaneously. The date of the avalanches is not well pinned down but spans the crucial time period of about 1,100 years ago.

Landscape-changing avalanches have been common in the Northwest. In fact, a series of avalanches are known to have converted one lake to two. As the ice age ended, melting snow filled glacier-carved canyons west of present-day Port Angeles to create what we know as Lake Crescent. At some point after the lake filled 10,000 years ago, landslides roared down steep mountain slopes and built a wide earth dam across the lake, dividing it in two; the lakes are known today as Lakes Crescent and Sutherland. There are 500-year-old trees growing atop the dam, so the slides must have occurred at least five centuries ago. Over the years, repeated slides built the dam between the lakes so high that Lake Crescent had to find a new outlet to the Strait of Juan de Fuca.

Hundreds of years ago there was a catastrophic landslide in the Columbia Gorge where the river cuts through the Cascade

Range. The mountainous north bank of the Columbia River collapsed, sliding across and temporarily damming the river. The slide might have been triggered by a small earthquake that jiggled loose an unstable slope, or by an unusually wet storm that saturated the slopes. The natural dam, more than 200 feet high, backed up the river as far as Hood River, Oregon. Early research showed that trees inundated by the floodwaters died about 700 years ago, but recent work indicates the massive slide may have been more recent, perhaps only 250 to 450 years ago.

Estimates based on average river flow tell us that it would have taken the Columbia about a year to rise high enough to cut a new channel through the blockage. The remains of the natural dam left a treacherous stretch of rocks and falls that pioneers named the Cascades. Although these dangerous rapids are now submerged beneath the Bonneville Dam reservoir, the river's course still swerves more than a mile to the south at that point, a reminder of the thunderous avalanche that pushed the river out of its ancient course and blocked the canyon. Local Indian tribes handed down a story about a long-ago time when it was possible to walk across the river at this spot. White settlers, dismissing it as legend (or perhaps just misinterpreting it), called it "the Bridge of the Gods" story.

Danger Ahead

What geological hazards does the future hold for the Pacific Northwest? A rule of thumb in geology is that whatever has happened before is almost sure to happen again.

Earthquake hazards

Every state and province in the Pacific Northwest has experienced powerful earthquakes. The largest earthquake in the region's recorded history occurred off the west coast of British Columbia's Queen Charlotte Islands in 1949, with a magnitude of 8.1.

Scientists believe a huge subduction earthquake occurred off the Washington-Oregon coast about 300 years ago, before white settlement and written record-keeping began in the area. Evidence includes radiocarbon dates on killed trees, tree-ring growth patterns, and historical accounts of a huge tsunami from the east that devastated the Japanese coast. Native Americans in the Pacific Northwest used to tell a story, dating back to the eighteenth century, of a huge seawave that destroyed coastal villages.

A quake that took place in the North Cascades in 1872 is generally considered Washington's biggest land-based temblor. Judging from reports in the still sparsely populated country, the quake was felt all over the Pacific Northwest; it collapsed buildings, caused landslides, and forced spurts of water from the ground. Geologists assume it was neither a subduction nor an intraplate quake but rather one that occurred at a fairly shallow depth in the crust. Its magnitude has been estimated at 7.4.

Oregon and Idaho have also experienced powerful earthquakes. Oregon's largest quake was in 1873. It was centered along the Oregon coast at the California border and had an estimated magnitude of 6.75. Areas in southern Idaho, southeast Oregon, and southwest Montana continue to be pulled apart along various fault lines. Deadly quakes in Montana in 1959 and Idaho in 1983 demonstrate that the faults are still active.

We can expect more powerful (but not catastrophic) earthquakes like those that shook the Puget Sound area in 1949, 1965, and 2001. And given the steep slopes in the western parts of British Columbia, Washington, and Oregon, there undoubtedly will continue to be landslides triggered by wet winters or possibly earthquakes.

The greatest earthquake hazard is along the outer coasts of Washington and Oregon, where a great subduction earthquake offshore (like one 300 years ago) would cause severe shaking and sudden subsidence. There could be as little as 20 minutes between the quake and the arrival along the coastline of a destructive tsunami. Destruction of inhabited areas would be complete. Farther inland, Portland, Seattle, and Vancouver, British Columbia, would be buffered from the worst effects of a great subduction quake but still could sustain extensive loss of life and severe damage.

Stresses of a different kind will make the ground tremble on both sides of the Cascades. These quakes, usually not strong enough to cause extensive damage, occur in the shallow crust and reflect north-south compression, as distinct from the northeast squeeze from the subducting Juan de Fuca plate. The cause of the north-south surface compression in the shallow crust is controversial.

Volcanic hazards

In the future, more volcanic eruptions in the Cascades are almost certain, with the most likely candidates being Mount St. Helens, Mount Rainier, and Mount Baker in Washington, Mount Hood in Oregon, and Lassen Peak in northern California. In the Cascades, lava flows have rarely traveled beyond the erupting mountain itself and are the least of the volcanic hazards. Mount St. Helens has been a prolific ash producer and in 1980 sent up an ash cloud that smothered downwind towns and eventually blew around the world. But the most serious danger from Pacific Northwest volcanoes will be mudflows. Mount Rainier appears to be the biggest threat: most North American volcanoes are remote from population areas, but valleys lead from Rainier's slopes directly to farms and dairies, and, not too much farther, to the Tacoma area.

Mount Rainier, the Pacific Northwest's highest and bulkiest mountain, is worrisomely unstable. Rock in the old volcano has been weakened by thousands of years of heat and corrosive volcanic gasses. Perched atop the rotten rock is a huge volume of ice and snow. Geologists say it wouldn't take much—perhaps a steam explosion or one of the small earthquakes that rattle the mountain periodically—to jar things loose. There wouldn't necessarily be warning signs before a torrent of steaming-hot mud, boulders, trees, and other debris thundered off the mountain into the valleys leading toward Puget Sound. It would be a human catastrophe of a scale unprecedented in this part of North America.

Giant mudflows have roared off Mount Rainier in the past. A mudflow from the mountain moved down the Puyallup River Valley about 600 years ago, leaving mud 15 feet thick on the site of present-day Orting, a suburb of Tacoma. Much earlier, around 5,700 years ago, a bigger flow was caused by a huge collapse that knocked an estimated 2,000 feet off the volcano's summit. The mudflow reached a valley (at that time an arm of Puget Sound) just south of present-day Seattle and today occupied by the towns of Auburn, Kent, Sumner, and Puyallup. Archeologists found tools beneath the deposit on a farm near Enumclaw, southeast of Seattle, proving that Native Americans were living in the area at the time the mudflow occurred. With the mudslide filling entire valleys, it is likely the local population suffered some loss of human life.

The threat of repeat mudflows prompted this warning in a U.S. Geological Survey report in 1992: "Mount Rainier volcano is potentially the most dangerous of the periodically active of the Cascade Range between Northern California and British Columbia." Scientists have placed automated instruments on Rainier in hopes they can send an alarm if something happens high on the mountain that presents a significant danger to the people below. But evacuating even a small city and getting people to high ground in less than an hour would be a formidable undertaking, particularly if the warning came in the middle of the night.

Far in the Future: A Changed Northwest

And how about the really, really distant future? If the North American plate continues its slow westward migration over enough millions of years, it will eventually override the undersea spreading ridge that is creating the Juan de Fuca plate. At that point, the entire western edge of the continent will be in direct contact with the Pacific plate. The San Andreas fault system will grind along uninterrupted from Mexico to the Aleutian Islands. There will be more frequent earthquakes along the Oregon-Washington coastal strip, but they will be those typical of California, rather than the great subduction quakes that apparently have shaken the Pacific Northwest in the past.

Deprived of magma rising from a subducting plate, the Cascade volcanoes will snuff out. The Cascades and the Coast Range in Oregon and Washington, including the Olympics, will no longer be squeezed higher; eventually, they will be eroded by rain and snow. Without high mountains, the climate in the eastern parts of Oregon and Washington will become wetter and warmer.

~

Although we're ending our account of how the Northwest was assembled, its building, tearing down, and shaping continues. Our story, which began in the distant past when our part of the globe was mostly open sea, ends in the present day when highways span old faults and cities dot the patchwork landscape made up of once widely separated terranes. The ground beneath us is restless.

❖❖❖

Pacific Northwest Geological Almanac

Largest earthquake in the Pacific Northwest

- ❖ Cascadia earthquake: estimated magnitude of 9.0 or above
  Colossal subduction quake believed to have occurred about 300 years ago off the Washington-Oregon coast, creating a tsunami that devastated the coast of Japan.

Magnitude of largest earthquakes, by state and province

- ❖ British Columbia: 8.1
  Off the coast of Queen Charlotte Islands, 1949.
- ❖ Idaho: 7.3
  Borah Peak, Lost River Range, 1983.
- ❖ Oregon: estimated 6.75
  Along the coast near the Oregon-California border, 1873.
- ❖ Washington: estimated 7.4
  North Cascades, 1872.

Most dangerous volcano

- ❖ Mount Rainier
  Risk of mudflows to inhabited area, especially Tacoma, Washington, and surrounding suburbs.

Volcano covering the greatest area

- ❖ Newberry Crater
  40 miles long and 20 miles wide, in southern Oregon.

Biggest ash eruption

❖ Mount Mazama
Eruption approximately 6,900 years ago in southern Oregon spread ash as far as Alberta, Canada, to the north and Nevada to the south.

Youngest volcano

❖ Mount St. Helens
40,000 years old.

Most recently active volcano

❖ Mount St. Helens
Active 1980 to 1986.

Highest mountain, by state and province

❖ Washington: Mount Rainier
14,411 feet, Cascade Range.
❖ British Columbia: Mount Fairweather
15,299 feet, British Columbia-Alaska border.
❖ Idaho: Borah Peak
12,662 feet, Lost River Range.
❖ Oregon: Mount Hood
11,245 feet, Cascade Range.

Deepest gorge

❖ Hells Canyon (Snake River)
Average depth: 6,000 feet from rim. Runs along the Idaho-Oregon and Idaho-Washington borders.

Oldest rocks

❖ Belt rocks, around one billion years old
Found in the area where Washington, Idaho, and British Columbia come together, extending into northwestern Montana.

Youngest rocks

❖ Recently formed rocks at Mount St. Helens
Some are only about 15 years old.

Longest river

❖ Columbia River
Flows 1,210 miles from the Rocky Mountains in British Columbia to the Pacific Ocean at the Washington-Oregon border.

Longest ocean-bottom channel

❖ Cascadia channel
Runs for 1,300 miles off the Washington-Oregon coast.

Largest lake

❖ Glacial Lake Missoula
530 cubic miles, approximately half the present volume of Lake Michigan.

Greatest floods

❖ Glacial Lake Missoula floods
Over 40 massive floods inundated parts of Washington approximately 13,000 to 15,000 years ago. Caused by the repeated draining of Glacial Lake Missoula.

Farthest advance of ice during last ice age

❖ 15 miles south of Olympia, Washington
Part of an ice sheet that extended 160 miles south of the Canadian border approximately 14,000 years ago.

Deepest ice during last ice age

❖ One mile deep, at the Canadian border
The ice sheet was only about half a mile deep over the future Seattle.

Least rapid transit

❖ One inch per year
Juan de Fuca and North American plates creep toward each other at the rate of one inch per year, making the collision velocity two inches per year.

Farthest-traveled lava flow

❖ Columbia River basalts in eastern Washington
Flowed 450 miles to the Pacific Ocean. Erupted along what is now the Idaho-Washington border.

Loftiest former ocean bottom on the Olympic Peninsula

❖ Mount Olympus
7,965 feet.

Year Los Angeles may reach present latitude of Seattle

❖ 50 million years from now
Assuming the Pacific and North American plates each move at the rate of one inch per year.

Glossary of Geological Terms

Andesitic lava. The thick, sticky, explosive type of lava that seldom flows far from the eruption site, often building high stratovolcanoes.

Asthenosphere. The region of the earth's mantle from 60 to 200 miles below the surface where rock becomes ductile and easily deformed.

Back-arc spreading. The stretching of the earth's crust that results in a thinned basin between a chain of volcanoes and a continent. The stretching can be caused by an oceanic plate (on the other side of the volcanoes) subducting faster than the continental plate that is moving toward it.

Basalt. A dark, fine-grained volcanic rock. The majority of rock in the Columbia Basin is basalt.

Basaltic lava. A very hot, fluid lava that may flow quickly and for some distance from the source, usually a fissure. It may form a broad, sloping shield volcano.

Batholith. A very large mass of volcanic rock formed from magma injected into older layers of rock; a pluton more than 40 square miles in area.

Caldera. A steep-walled basin created when a volcano collapses into an emptied magma chamber.

Cinder cone. A cone built up of small pieces of basalt ejected from a volcano. The cone may reach several hundred feet in height before it produces a small lava flow.

Continental drift. An old name for the idea that continents drift across the earth's surface, now known as plate tectonics.

Continental slope. The slope from the coastline to deeper ocean water; the true edge of a continent.

Crust. The rigid outer layer of the earth's surface, extending to a depth of about 30 miles.

Fault. A fracture in the earth's crust along which movement has occurred.

Hot spot. A fixed spot in the earth's mantle that periodically sends lava to the surface of a tectonic plate moving above it, creating a chain of volcanoes.

Intraplate earthquake. An earthquake caused by movement within a subducting plate rather than by movement of the plate itself.

Lava. Molten rock that has erupted onto the earth's surface.

Lithosphere. The outer layer of the earth's surface, extending down about 60 miles, composed of rock lighter and more rigid than rock in the asthenosphere. Includes the crust.

Magma. Molten rock beneath the earth's surface; once it reaches the surface, it is called lava.

Magnetic reversal. Reversal of the earth's magnetic field that has occurred repeatedly in the distant past for reasons that are not entirely clear, most recently about 700,000 years ago.

Magnetometer. A device that when towed by a ship detects faint magnetic forces in the rock below the ocean floor.

Mantle. The thick shell of rocky material between the earth's crust and the core.

Obsidian. A glassy volcanic rock that cools without forming crystals.

Paleomagnetism. Magnetic clues in volcanic rock recording the direction of the magnetic poles and the approximate latitude of the rock at the time the lava cooled.

Plate. A section of the earth's crust that moves as a unit relative to other sections.

Plate tectonics. The science that studies the movement of segments of the earth's crust.

Pluton. A mass of volcanic rock formed from magma injected into older layers of rock.

Radiocarbon dating. Using the known decay rate of a radioactive form of carbon to determine the age of a sample of once-living material.

Right-lateral fault. A strike-slip fault in which, from an observer's standpoint, the relative motion of the opposite side of the fault is to the right.

Seafloor spreading. The lateral movement of sections of seafloor away from the mid-ocean volcanic ridges where they were formed.

Seamount. A volcanic mountain that rises several thousand feet above the deep-sea floor but does not reach the ocean surface.

Shield volcano. A volcano that erupts fluid lava and forms a broad, dome-shaped mountain with gentle slopes.

Spreading center, or spreading ridge. The mid-ocean ridge where segments of seafloor move away from each other.

Stratovolcano. A steep-sided volcano built of layers of lava, ash, and rock fragments, such as Mount Rainier in Washington.

Strike-slip fault. A fault in which the earth movement is horizontal and parallel to the direction of the fault.

Subduction. A process in which a heavy ocean plate dives beneath lighter continental material.

Subduction zone. The boundary between oceanic and continental plates where subduction takes place.

Superterrane. A number of terranes pushed together by plate movement.

Terrane. A segment of the earth's crust that has a distinct geologic character as compared to neighboring terranes.

Tsunami. A very large and often destructive ocean wave caused by an undersea earthquake.

Turbidity current. A dense current that moves along the ocean floor, carrying sand and rock particles dislodged from the continental slope.

Index

XYZ

About the Author

Photo by Roy Scully

Hill Williams was a science writer for *The Seattle Times* for twenty-four years. He received a bachelor's degree in journalism and a master's degree in communications from the University of Washington. He has been the recipient of numerous regional excellence-in-journalism awards and in 1984 received the American Association for the Advancement of Science-Westinghouse Science Journalism Award (also known as the AAAS-Westinghouse Award), given for distinguished science writing in newspapers with over 100,000 in daily circulation. Mr. Williams lives in Shoreline, Washington.